# ROUTLEDGE LIBRARY EDITIONS: HUMAN GEOGRAPHY

Volume 18

# GEOGRAPHY AND POLITICAL POWER

# GEOGRAPHY AND POLITICAL POWER

## The Geography of Nations and States

PETER M. SLOWE

Routledge
Taylor & Francis Group

LONDON AND NEW YORK

First published in 1990 by Routledge

This edition first published in 2016
by Routledge
2 Park Square, Milton Park, Abingdon, Oxon OX14 4RN

and by Routledge
711 Third Avenue, New York, NY 10017

*Routledge is an imprint of the Taylor & Francis Group, an informa business*

*British Library Cataloguing in Publication Data*
A catalogue record for this book is available from the British Library

ISBN: 978-1-138-95340-6 (Set)
ISBN: 978-1-315-65887-2 (Set) (ebk)
ISBN: 978-1-138-95728-2 (Volume 18) (hbk)
ISBN: 978-1-315-66174-2 (Volume 18) (ebk)

**Publisher's Note**
The publisher has gone to great lengths to ensure the quality of this reprint but points out that some imperfections in the original copies may be apparent.

**Disclaimer**
The publisher has made every effort to trace copyright holders and would welcome correspondence from those they have been unable to trace.

# Geography and Political Power

The Geography of Nations and States

Peter M. Slowe

Routledge

London and New York

First published in 1990 by Routledge
11 New Fetter Lane, London EC4P 4EE

Simultaneously published in the USA and Canada
by Routledge
a division of Routledge, Chapman and Hall, Inc.
29 West 35th Street, New York, NY 10001

Typeset by Mayhew Typesetting, Bristol
Printed in Great Britain by
Richard Clay Ltd, Bungay, Suffolk

*British Library Cataloguing in Publication Data*

Slowe, Peter M.
  Geography and political power : the geography of
  nations and states.
  1. Political geography
  I. Title
  320.1'2

ISBN 0-415-03912-6

*Library of Congress Cataloging in Publication Data*

Slowe, Peter M.
  Geography and political power : the geography of
nations and states / Peter M. Slowe.
    p.  cm.
  Bibliography: p.
  Includes index.
  ISBN 0-415-03912-6
  1. Geopolitics. 2. Power (Social sciences) 3. World
  politics. I. Title.
JC319.S56    1990                          89-10399
320.1'2-dc20                                    CIP

This book is dedicated to the memory of my father
**MALCOLM SLOWE**

# Contents

# List of figures

# Acknowledgements

I would like to thank all the friends and colleagues who have helped me prepare this book.

In particular I am grateful to Ian Dutt for introducing me to Hip Hop, to Lyndon Gainham for his Channel Islands survey, and to Andy Lane for his ideas about the Western Sahara. I am also very grateful to my wife Karen and our friend Nikki Sparks who shared the typing, to my mother who kindly helped with my research on Guinea, to Diana Smith the excellent cartographer, and to the very helpful staff of the library at the West Sussex Institute of Higher Education.

My wife Karen was patient and tolerant throughout the occasional takeover of our household by *Geography and Political Power* and I could not have written the book without her being around – and producing our son Alistair at the end of Chapter Four! My late father, to whom this book is rightly dedicated, would have been rather pleased with both the baby and the book.

Peter M. Slowe
Angmering, West Sussex

# General introduction

Geographers are more important than they think. The subject they study lies at the very heart of political decision-making. From the World Bank to the parish council, from superpowers imposing their will on their neighbours to a provincial independence struggle in a famine-struck African state, it is geography that determines economic and political strategy and geography that determines cultural and military tactics. Only geography provides complete models of the world in which decisions are made; only geography focuses on the relationships between culture, economy, society, and polity, to model the whole landscape. The economist's landscape is only economic, the historian's landscape is only historical, the political scientist's landscape is only political, and the sociologist's landscape is only social; the geographer's landscape is the real landscape of the decision-maker and the real landscape of action, because it is a landscape in which the objects and relationships which matter are highlighted, irrespective of the discipline into which they might otherwise fall; geography is the science of the land and the whole human environment. The problem is that geographers have forgotten the significance of what they do. Earlier in this century, and in the last century, geographers never hesitated to pronounce on all the political issues of the day; the work of geographers like Ratzel and Mackinder, for example, was central to political affairs, but recently geographers have allowed themselves to be pushed into the political sidelines and away from national and international affairs where they have the skills to contribute to matters of peace and war.

The aim of this book is to show the connections and introduce the modes of thought that will help to propel geographers back to the centre stage of national and international politics. Since it is anyway the case that geographical modes of thought are used in political decisions, what is now needed, if geographers are to reassert themselves, is for them to be aware of the political philosophies that lie behind the five main sources of political power. These are 'might' (aggression and conquest), 'right' (the mobilization of people who feel they have a right to territory), 'nationhood' (the unique power of the nation-state), 'legality' (the legal distribution of power and territory) and 'legitimacy' (the reality of hegemonic power within and between states). The study of these five sources of power goes deeper into an understanding of politics than the geographical adaptation of some political ideas which is traditionally called 'geopolitics'. This book discusses geopolitics but only to provide in certain cases a clearer understanding of the implications for geography of political thought.

Each of the book's five chapters explores one of the five sources of power. In every case, there is an opening section on the general political philosophy associated with the particular source of political power. It is always followed by a section which develops or highlights some important aspect of the philosophy. Case-studies then illustrate the areas where the political philosophy is applied; each case illustrates a field where geographers have an important contribution to make.

In Chapter One on 'Power through Might', part one on philosophy is followed by a study of the psychology of power through might. The case-studies are of Nazi Germany, the Amazonas, South Africa and Namibia, and the Western Front in the First World War. These illustrate the geographical consequences of the aggressive use of power by individuals, by states, and by groups within states.

The use of brute strength to change the political world is as old as history itself. It is a good starting-point for a book which sets out to illustrate the relationship between geography and political power, because it is so obvious, simple, and central in political geography. It is rarely possible to divorce the assertion of brute strength from concepts such as right and nationhood, but it is usually feasible to identify a feeling of being empowered through armed might as a prime cause of certain battles, campaigns, wars, and so on. There is also a philosophy which is orientated around the concept of might, which is relevant to a good deal of political geography.

The philosophy of power through might is discussed in the first part of Chapter One. There are two main strands in this philosophy: the philosophy of the assertion of the individual within the state and the philosophy of the assertion of the power of a particular state within the community of states. Both these philosophies of assertion, however, would be meaningless unless individuals on the one hand or states on the other were aggressive, so the next section of Chapter One discusses aggression. The philosophy of power through might provides the framework for thought and this combines with human aggression to cause war. The potential for aggression lies within every individual and, when individuals combine into groups, aggression starts to have a significant influence on political geography. It is the aggression of groups – races, nations, armies, and so on – that provides sides to be taken and enables wars to be started.

The first case-study of power through might is perhaps the most obvious case of planned aggressive war: Nazi Germany. Fortified by a form of geopolitics which could trace its origins to the philosophy of power through might, Nazism provided one of the clearest models in political geography, stating simply that greater strength leads to a right of conquest. Hitler's personal opinions gave a racial twist to the latest stages of Nazism's aggressive war, but the basic model of Nazism, the assertion of armed strength, lasted from the writing of *Mein Kampf* until the fall of Berlin in 1945.

The second and third case-studies, the Amazonas and South Africa's involvement in Namibia, are further examples of the assertion of power through brute strength, backed up by various geopolitical notions such as the need to occupy empty or potentially valuable areas in Latin America and the need to maintain a certain life-style for Europeans in South Africa.

The final case-study in Chapter One takes account of the fact that once it has come to war, whatever the reason and whatever the history, every dispute is reduced to a test of power through might in which carefully channelled aggressive group behaviour is the norm. The battlefield is where the gain or loss of territory is finally determined, and the battles are won by people motivated to aggression by a variety of emotions including sheer excitement. Assertive man spoils for aggressive war and enjoys the fight; this simple sequence is at the heart of power through might. The battlefield is the real and symbolic focus of many an individual's and many a state's hopes and dreams, and this is illustrated in case-study 1.4.

Although the assertion of strength lies at the heart of many wars and all wars are resolved by trials of strength, an alternative reason for a war might be the assertion of a sense of right. This is the focus of Chapter Two, 'Power through Right'.

In Chapter Two on power through right, the first section on philosophy is followed by a study of the development of the territorial state which has become the main focus for groups of people claiming a right to territory in the modern world. The case-studies are of the Holy Land, the Falklands, Grenada, and Northern Ireland. They illustrate the mobilizing effect of the man–land relationship which makes people fight – sometimes literally – for the land they believe they should occupy or own.

The philosophy of power through right is concerned with the ownership of land and attachment to land. Compared with power through might, it is concerned less with armed strength than with emotional strength. It is the feeling of a right of ownership and a sense of belonging which motivates people to argue and fight for territory which they consider to be theirs.

The traditional philosophical debate has been between those who consider the ownership of property, particularly territory, to be natural or unnatural. This theoretical argument can be eclipsed by referring to the empirical literature on the behaviour of children and people in primitive societies and their view of territory; the source of power through right is shown to be a powerful human emotion (the origin of which through nature or nurture is of little interest, since the two are not really distinguishable in these cases – and probably not in other cases either).

The second part of Chapter Two explains why the territorial state is the most important focus of attachment to territory and of group senses of belonging; a man is more likely to die for the state than for any other territory greater or smaller. For economic and cultural reasons, the territorial state replaced the great medieval dynastic and religious realms. Above all, modern economies and societies need something like a state framework within which to operate: they need the state's administration of law, order, defence, and security and its positive functions such as welfare and education. Because the state is so important, it is the main focus of power through right.

The first case-study of Chapter Two is of the Holy Land. Just as Nazi Germany was an obvious example of power through might, so the Holy Land is the classic case of power through right. Jews and Arabs equally claim the right of ownership and each feels a strong sense of belonging. Each is willing to fight if necessary against the odds for the Holy City of Jerusalem, for the biblical Israel of King David, for the modern state of Israel, and for the West Bank. Such is the conflict over the right to land, rooted in culture, politics, and theology.

The second case-study has led to only one war, but feelings run almost as high as in the Holy Land. The Falklands War of 1982 is explained in terms of the conflicting feelings of right between the British claim, based on self-determination, and the Argentine right, based on the idea of the thwarted destiny of a would-be South Atlantic power. The Argentine claim also illustrates the importance of physical geographical nearness in generating a claim of a right to run territory.

A feeling of the right to self-determination is the aspect of power through right considered in the third case-study of Chapter Two. This is the case of Grenada. The case illustrates the real limitations on self-determination. Grenada is a nominally independent state but in reality its independence is severely limited by falling within the sphere of influence of the United States. Possibly real self-determination is enjoyed to a greater extent with the protection of a major power, such as in Martinique, an overseas *département* of France, or the Falklands under Britain. Grenada only appears to have had nominal independence. The feeling of a right to self-determination is powerfully motivating, but nominal political independence and actual self-determination are two very different things, as Grenada discovered at the cost of some bloodshed.

The final case-study, Northern Ireland, focuses on possible resolutions to conflict over territory in the light of the failure there of the parliamentary system. Possible solutions include uniting the island of Ireland and the systematic unmixing of groups in conflict.

If might is strength through brute force and right is strength through emotion, nationhood is strength through a subtle combination of the two, reified by the most durable and widely recognized territorial entity in the world today, the nation-state. It is argued that nationhood is in itself, even without armies and even without strong emotional attachment, a source of strength, especially when it is combined with the territorial state.

In Chapter Three on 'Power through Nationhood', part one on philosophy is followed by a section specifically on the way in which the sovereign state generates the affinities and transactions of nationhood. The case-studies are: England after the Norman Conquest; pre-colonial nationhood in Guinea; integration and nationhood in Touré's Guinea; and disintegration, reintegration, and nationhood in Conté's Guinea. These illustrate the development in one territory of both nationhood and statehood and, thus, of the single most powerful political entity in today's world and the one that most directly affects people's lives: the nation-state.

The philosophy of power through nationhood comes to grips with definitional problems as well as sorting out the complex inter-relationships between ethnicity, nationhood, and state. These three concepts are open to a variety of interpretation. Here they are used quite specifically. 'Nationhood' is the same as 'ethnicity' except that it occurs within state boundaries and is thus created or reinforced by the power of the state. Nationhood without the state is therefore impossible, although there may be some aspiration to a state or 'nationalism'.

Ethnicity is the consequence of affinities and transactions outside the framework of the state among a group of people coherent enough to be called an ethnic group. Similar affinities and transactions to those which characterize an ethnic group are also actually generated by the state itself, for the state is extremely powerful, deriving its power from its sovereignty over clearly defined territory, its universal compulsory and sole jurisdiction within definite boundaries. The state can easily generate the affinities and transactions which occur naturally within an ethnic group and, when it does so, the consequence is nationhood.

Not every state generates nationhood; but every state aspires to do so because nationhood provides the state with a consensus and thus with the power to mobilize and motivate citizens and to obtain their loyalty. The second part of Chapter Three discusses the state.

The state, the creator of the nation, is distinct from every other sort of organization. It has universal, compulsory jurisdiction within its territorial boundaries. It has sovereign authority, recognized by other states and by its own people. This power is reinforced by the very uniqueness of the state. So much of human life is carried on within it, through it, and by it that inevitably it contributes to the set of affinities and transactions which make up the nation, incorporating through simple territorial boundaries those who might even have been excluded by ethnic groups before the state came into being. In effect, the state creates its own ethnicity which is called 'nationhood'.

The first case-study is of the unplanned development of nationhood in an embryonic state: England under the Normans. Feudalism played a key part, as it evolved, in creating an English nation from two very distinct ethnic groups in one state territory. English nationhood became a source of unity and therefore a source of strength in the future.

The last three parts of Chapter Three form a single study of the West African state of Guinea. For the purposes of this study, a new concept is introduced, 'integration', including all kinds of interactions for trade, society, and culture. Integration intensifies the transactions and affinities which create both ethnicity and nationhood; in the case of nationhood, integration may enable the state to become a nation-state.

Integrated pre-colonial empires gave way to economic and social policies of disintegration and division in the colonial era. From independence in 1958 until 1984, Sekou Touré's Parti Democratique de Guinée often sacrificed more traditional forms of development for integration. Through policies for integration, Guinean nationhood inevitably evolved and Guinea became a nation-state, a relatively united entity, with fewer class divisions, ethnic divisions, or spatial inequalities than most other African states. This gave Guinea the strength to survive without serious dislocation a change of regime to one unconcerned with integration and nationhood: the new regime reopened Guinea to the west in a sudden and dramatic way.

Consciously developed nationhood has given Guinea the stability and strength to survive the inpouring western capital, capitalism, and capitalists.

The state reifies the nation and the state itself is defined by boundaries. These boundaries are legal entities, recognized by international courts and defined by constitutions. They literally circumscribe the power of the state, making its territory great or small, defensible or indefensible, and sometimes even rich or poor. They are a major source of power and are the focus of Chapter Four.

In Chapter Four on 'Power through Legality', the section on philosophy is followed by a study of the state boundary, which is the direct focus of many disputes and wars between states, even though the original causes may be more profound. The case-studies are of the Sino–Soviet boundary dispute, a federal boundary (between Britain and the Channel Islands), Western Sahara, and the end of the partition of Jerusalem. These illustrate the important two-way relationships between the geography of the boundary and the politics of the state and between the politics of the boundary and the geography of the state. Boundaries, a traditional focus for geographers, affect the distribution of power and affect numerous people's lives.

One of the main problems of the geography of state boundaries has been the absence of a theoretical framework. This has limited the insights which such studies have been able to provide into the wider political geography of the bounded territorial state. The first part of Chapter Four, on the philosophy of power through legality, examines the important work of Sven Tagil and others, which provides a way of understanding how boundaries function. Tagil's approach can be applied to any boundary analysis and effectively highlights salient points. These may generally be defined as dominant values or primary objectives from either side of a boundary and, if there is a special incident or development, there may also be particular actors or 'situational' objectives to consider, suggesting various forms of action. For additional insight into boundary disputes, the Tagil model can be used alongside political conflict theory. More generally, the Tagil model can complement geographical models of the delimitation and administration of boundaries and barriers (discussed in part two of Chapter Four).

The first case-study is the Sino–Soviet boundary dispute, which is traced historically from the distant encounter of two medieval empires through to the Gorbachev–Deng reconciliation. The Tagil model is used explicitly, providing insight into conflicting general and specific values and with conflicting general and situational objectives.

7

The second case-study makes explicit use of the geographical insights in Chapter Four, in its study of the federal boundary between Britain and the Channel Islands. Factors strengthening and weakening the boundary are identified.

The third case-study is of the Western Sahara. This is a study of the relationship between boundary and state. A political territory was defined by the Spanish colonial power and had three conflicting claimants by the time of its independence, each drawing new boundaries and seeking to destroy old ones. Boundaries have been drawn and redrawn between the Western Sahara and Morocco and between Mauritania and Morocco across the Western Sahara. Saharawi independence was helped by the proximity of Mauritania's valuable phosphate industry to its vulnerable international boundary and hindered by the creation of a closed walled boundary built across the desert by the Moroccan occupation army. The problem may have been resolved by wider considerations in the politics of North Africa.

The final case-study is of a boundary which formally disappeared in 1967: the partition boundary between Israeli and Arab Jerusalem. The focus of this study is the boundary as a barrier. The physical and economic barriers disappeared but functional and social barriers remained.

This chapter on 'Power through Legality' still leaves one significant gap in the study of Geography and Political Power, not accounted for here or in preceding chapters. The hegemonic distribution of power at the world level, the state level, and at the level of the town or city, plays an important part in explaining geographical reality. It is this distribution of power through legitimacy – so called because most people accept it and therefore legitimate it – that is probably the most important source of power in terms of its effect on people's everyday lives. It is the focus of Chapter Five.

In Chapter Five on 'Power through Legitimacy', the section on philosophy is followed by a study of the world hegemonic system, neo-imperialism. The case-study of the operation of neo-imperialism in Ethiopia is followed by case-studies of hegemony at other levels, these being hegemony and town planning, Hip Hop, and Albania. The four case-studies taken together illustrate the effects of legitimated hegemonic power on the geography of the world.

The philosophy of power through legitimacy was developed by Gramsci from the ideas of Hegel and later refined by Galbraith. Hegemony is taken to be the domination of one small group over the rest of the population. For Hegel, those who were dominant were imbued by the *Geist*; for Gramsci it was the bourgeoisie; for Galbraith it was the technostructure, the managerial class.

The second part of Chapter Five shows how legitimated power operates at the world level in a system of neo-imperialism dominated by international economic power, overriding state boundaries. The example of agribusiness is discussed.

Applied to one state in the first case-study, Ethiopia's poverty and famines are explained in terms of the hegemonic domination of the world economy by a handful of international capitalists. The case of Ethiopia also demonstrates the tension in a hegemonic system of world capitalism between international capital and the state.

The next two case-studies look at the geography of the city resulting from the hegemonic distribution of power within the state. Town planning in the western city is shown to be increasingly obsessed by order and the ability to control. From Hausmann's Paris to the carefully designed suburban estate in the modern American town, the purpose is to provide the physical framework for a subdued society. The exciting alternative offered in Sennett's *The Uses of Disorder* is discussed briefly. The second of these two case-studies of the city is concerned with the cultural character of the inner city. It describes the tension in English cities between Hip Hop culture, which has enabled black youth to find a new identity, and an establishment which feels threatened by the only group to have rioted seriously in Thatcher's Britain.

The final case-study is of Albania, which stands out as an example of a state which has followed its own special development path. The Albanian Communist Party (Party of Labour) sees itself as guiding the dictatorship of the proletariat following a Marxist revolution; it sees it as its responsibility to impose its cultural and economic hegemony to speed the inevitable evolution towards Communism. The power of the Party is accepted – legitimated – by a population subjected to intense and unsubtle propaganda; but the Communist values imposed are new and suggest different ways of assessing culture, economy, and society.

Power is always contested. It inevitably involves disputes about how it should be used. These disputes are the subject matter of politics. The exercise of power influences the location and configuration of human activity, thus politics affects geography. This in turn affects the subsequent round of political activity. Thus geography and politics, or more especially geography and political power, are interrelated inextricably.

If power enters into the relationship between two people, or two groups such as states, it inevitably affects the character of both parties involved. One party usually takes precedence over another in some way, by legal authority or by authority achieved by force, manipulation, or exploitation. Legal authority involves all the various means by which popular compliance is attained in the world, including representative government and the popular acceptance of hierarchies such as families and clans of people or states, and in this way it gives one party the power to do things which affect other parties. Authority obtained by force, manipulation, or exploitation gives one party power over the other regardless even of theoretical popular acceptance. Power, by whatever sort of authority it is derived, affects people's lives. Geographers can do much to help everyone to understand power better and to use it more benevolently.

# Chapter one

# Power through might

## Introduction

The aggressive use of brute force for conquest and domination is an important cause of change in the world's geography. Political geography is obviously affected, but authoritarian governments and aggressive states have profound effects on economic and social geography as well. At the same time, geography itself influences those who take political decisions, and so a two-way relationship evolves. The aim of this chapter is to unravel the relationship between geography and political power which is based on force: 'power through might'.

First, the chapter introduces the philosophy of power through might, which traces the thought which culminated in German geopolitical ideas and ultimately in Hitler's doctrine of territory. Machiavelli, the philosophers of absolutism, and Utopian philosophers represent three traditions which, when applied, result in the domination of the state by one person or one elite group. Nietzsche, Treitschke, and the philosophers of Fascism attribute a personality to the state itself and it was this philosophy which lay behind the idea of a dominant state and behind ultimate German domination in Europe.

The second part of the chapter links philosophy to action through the philosophy of assertion applied to individuals and groups. There are three fields to consider. First, man's individual aggression is important for there could be no conquest and no domination without it. Second, man's aggression is all the more effective when he is a member of a group, such as a national or racial group. Third, a desire for power for its own sake, a desire for security, and a romantic yearning for glory are all shown to be crucial in turning ambitions and temperament into the battles and victories of aggressive wars. These are the wars without morality, the wars in which power is asserted through brute force.

Four main themes can be discerned where geography is related to

11

the philosophy of power through might and the psychology of assertion. The first of these is geopolitics and Hitlerism; the case of Nazi
Germany (case-study 1.1) looks at the role of geographers in the
development of Nazi thought and the practice and consequence of
Hitler's geopolitics. The second theme is the more general aim of
controlling territory for purposes of security and status. This is illustrated with reference to the Amazonas (case-study 1.2), including
Brazilian geopolitics in theory and also in practice in Brazil's internal frontier on the Amazon, and the importance of the empty
Amazon Triangle to Ecuador and Peru which have twice been to war
over it. The third theme arises from the current debate between
socialists and neo-conservatives about the domination by an elite
group in a state over its wider population, and there is no better
example of indisputable elite domination than South Africa and
Namibia (case-study 1.3). This includes the origins, theology, and
economics of the apartheid system of white domination and its effect
on Namibia. The fourth theme is individual and group aggression
and this has obvious consequences in battle: the Western Front in the
First World War (case-study 1.4) looks at methods of gaining
territory in battle, especially in waves of attack, and at the way
many people enjoy the fight. It also highlights how places where
battles were fought and lives were lost become shrines for
pilgrimage and memorials to glory in subsequent years.

Overall, this chapter shows that the assertion of power by the
simple use of superior strength is a complex phenomenon rooted in
a long history of philosophy and requiring a deep understanding of
human psychology. This aggression has important effects on human
geography, and geographical thought has profoundly influenced the
politics of assertion.

## The philosophy of power through might

### Individual power within the state

Machiavelli is a good starting-point for looking at the philosophy of
power through might. It is the ideas he proposed in *The Prince* at
the end of the fifteenth century that lie at the basis of the concept
of statecraft as a means of maintaining a central authority not
necessarily by consent but rather by guile or force (Machiavelli
1961). *The Prince* was written at a time when the need for a strong
central authority was felt by a rising bourgeoisie, and it was a time
when the apparently natural morality of medieval ecclesiastical thinking was no longer universally accepted. It was a time which called
for a kind of political realism where force is the essential

characteristic of the state and the basis for all action. The accumulation of the means to use force and the actual use of force were therefore the fundamental values of politics. The new spirit of Machiavelli's ruling class was enforcement, the enforcement of the will of the minority over the majority. People were either obstacles or instruments. These were not the values of the common man or of the peasantry, who were merely the masses to be inspired by a morality and by a religion to become useful instruments of authority.

Machiavelli was the first to link the existence of a system of states to political prudence. At a time when states were few and very fragile, their rulers were keen to develop the idea of the sovereign state that recognized no limits to its power within its boundaries. The ruler sits firmly at the peak of the hierarchy which overtly puts order before justice, since justice and humanity are only possible when essential authority is established – and it is not clear how complete that authority has to be. Order is the first concern of the state. The ruler is a supreme legislator above all laws. Law is first and foremost the instrument of authority of the ruler. The only limit to that authority is the territorial boundary of the state.

In the following two centuries, Calvin, Bodin, and Hobbes added some significant thought to the philosophy of state power. Calvin said relatively little about political structure, but his *Institutes of the Christian Religion* were important for their great influence and their emphasis on the citizen's duty to obey while having certain liberties which should be respected by any wise authority (Calvin 1980). Bodin's *Six Livres de la République* (1961) discussed further the natural liberty of the citizen but limited it still in the Machiavellian tradition of strong, unchallengeable, monarchical authority. Bodin was the strongest advocate of enlightened absolutism; absolutism limited by the natural freedom of individuals or self-limiting in the light of the implicit aim of keeping citizens happy. Hobbes's ruler was extremely powerful, but in both *De Cive* and *Leviathan* he conceded great new personal liberties to citizens. These included religious freedom, educational freedom, and a whole range of liberties likely to enhance trade, peace, and good order (Hobbes 1949, 1968). All this contrasts with the more demanding totalitarian states advocated by, or implied by, influential Utopian philosophers.

Individuals play no part in Plato's Utopia, a grim *Republic* with a strictly hierarchical and oligarchic rule; and Plato was the model for a school of philosophy which longed for an imposed regimented state (Plato 1955). Morelly in his *Essay on the Human Spirit* (de Foigny 1952), for example, noted that private property led to greed and misery, therefore private property should be abolished. Morelly planned a paradise, enforced by drastic penal laws, in which every unit of work and every individual had a controlled place. Saint-Simon, originator of the Saint-Simonian communes, backed this up with the argument that progress in the form of welfare, efficiency, prosperity, and the elimination of poverty can only be brought about by a society headed by men of industry and science who will, by nature, be decent and not tyrannical; any form of democracy is irrelevant (Manuel 1965).

As firmly in the totalitarian tradition as Plato, Morelly, and Saint-Simon were (if unwittingly) the power-from-below Utopians such as Rousseau, Hegel, and Marx. Admittedly, the desire to force history and geography into shape was not so overt, but it amounted in practice to much the same thing.

Rousseau declared that liberty consists less in doing one's own will than in not being subject to the will of another; yet his obsession with the 'general will' (which represents what people actually want rather than what they think they want, and which can only be understood by a great ruler) leads inevitably to totalitarianism – for example the enforcement of a new religion (Rousseau 1968a). Hegel's *Philosophy of Right* leads the same way by another path; Hegel gave the state primacy over individuals; he argued that in the end individuals will wish to surrender their liberty because the legal status of one man is interwoven with the rights of everyone else, and so individuality is best subsumed to the will of an undemocratic and very powerful state with just a few safeguards such as the guaranteeing of some religious freedom (Hegel 1942). Arguably, Marx follows a similar line when he alludes to post-revolutionary administration, a kind of arbitrary rule in which safeguards will no longer be necessary because man will have reached a higher stage of society (Marx 1981, vol. 3).

Three main trends of thought are discernible in the discussion so far: absolutism, Utopia imposed from above, and Utopia inevitably achieved by ordinary individuals acting together in the collective interest. Three influential sets of philosophies lent support to the idea of rulers asserting power over the ruled. Machiavelli and Hegel in different ways placed this assertion in the context of the state; Machiavelli laid the ground rule that the state is the most effective sphere of operations for the assertive ruler; and Hegel lent the state a mystical supremacy over the individuals who live in it. In the nineteenth and twentieth centuries, this idea of the state developed until it inevitably spilt over into the analogy of state for individual; thus, before, individuals were naturally suppressed by other assertive individuals but, now, states would be equally naturally suppressed by other states as well.

State power within the community of states

The state was transformed by nineteenth-century philosophers like Nietzsche and Treitschke from something that was primarily practical into a cultural world theory. The state was a superior entity for which sacrifice was right and proper. There was a contempt for concepts like compassion or pity which might inhibit the assertion of the state over individuals; the weak and inferior had to go to the wall for the sake of the state. Nietzsche's superman had the power to keep the masses in order, to ensure a self-sufficient state with enough people and enough land, and a will to ensure that everyone fulfilled his obligation to the state. The state was above the struggles of society, a tangible power demanding service and sacrifice in return for protection and glory (Nietzsche 1961; Treitschke 1963).

It was the philosophers of Fascism, principally Gentile, d'Annunzio, Michels, and le Bon, who refined the philosophy of the twentieth-century assertive state. They built into their analyses three principal concepts: popular culture, natural law, and self-fulfilment through the state.

First, popular culture included the instincts that people were supposed to have in favour of national traditions, family bonds, and a sense of belonging to definite groups (particularly if the latter were associated with some idea of traditional morality). Popular culture was about positive action rather than dogma, preferring liturgical catch-phrase to discussion. The philosophers of Fascism built these crowd instincts into their political philosophy. In German philosophy, they added a kind of overawed but passive appreciation of past creativity; there was something of the tradition of Kant and Goethe in the idolization of past cultural works as if they were divine revelations. This approach was later developed more or less to absurdity in Spain by the Fascist philosopher Primo de Rivera. According to Fascist theory, popular culture is shared among all non-deviant citizens. They are a *Volk* with a common goal. They call for government with a leader who has activated their will through his own energy.

Second, Arendt (1967) identified the crux of Fascist philosophy as being the way it subsumed every individual interest to natural law. Fascism made a conscious break with positive law in favour of natural law. The discrepancy between legality and justice could thus be neatly overcome, not by divinity as in the past, but by nature and history. If man were subject to overwhelming natural or historical processes which made him subject to his state and his leader, then the whole field of human law-making would be reduced to a matter of convenient regulations for day-to-day life.

Third, perhaps the most effective point of Fascist philosophy was that man, alienated in bourgeois society, could only fulfil himself through the state. Man was robbed of his individuality in industrial society. Fascism could recapture the whole man, channelling the individual will to the service of the state which in turn expressed, through its leadership, what citizens were really feeling and gave its citizens what they really wanted. They could not express these things for themselves because they were suppressed, alienated, and atomized, but the state could express it for them and thus indirectly provide them with fulfilment (Fermi 1961; Gregor 1969; Michels 1915; Thomas 1972).

### Conclusion

From medieval absolutism through Utopianism and the development of the Fascist state, a philosophy has developed which justifies the use of naked force by individuals over other individuals, by minorities over majorities, by one leader over all, by state over civil society, and by one state over others. It is a philosophy with a very simple view of people and society. It is a philosophy of aggression, force, and strength. It is a philosophy of assertion.

## The psychology of power through might

### Individual aggression

All the philosophy of assertion and aggression is as a grain of sand in the desert if man is not – or, at least, if some men are not – naturally assertive and aggressive. Is assertion natural? Is aggression natural? Is violence natural? If so, the psychology will combine with the philosophy to produce a violent world. Violence, aggression, and war may most commonly have their background in clear causes, and these are discussed in subsequent chapters. But there are also human traits, which are the concern of this section, which specifically develop the philosophy of aggression into acts of violence and war.

St Augustine, Luther, and Swift had no doubt about the natural evil of man – his violent aggression was symptomatic of the Fall. Their views, typical of their time, were firmly based on strong biblical belief rather than conventional argument (Augustine 1961; Ebeling 1970; Swift 1939).

It is probably more helpful to see man as a bizarre mixture of passion and reason, following Spinoza. Man is just a bundle of virtues and vices to be worked on and moulded by a variety of forces (Hampshire 1951). Marx, Morelly, Plato, Rousseau, and Saint-Simon would all happily go along with this idea. Man can be changed and perfected, and war can then be avoided for ever.

Great psychologists have gone down the same path as the Utopians. People just need treatment. Strachey in *The Unconscious Motives of War* (1957) said the hierarchical pattern of family social life frustrated children, who consequently learnt aggression at an early age, ripening them for war. Kluckhohn (1950) wrote about the central problem of peace as being a need to curb individual aggressive impulses, and Allport (1955) – writing at the height of the Cold War – implied that surveys should be conducted among the Russian population by American psychologists to see what was making the Soviet Union tick. Psychologists like Miller (1970) and political scientists like Deutsch (1966) argued that war arose out of ignorance and fear, and therefore that it really was only a matter of mental health and general knowledge to cure the world of war. All this was to ignore the fact that the causes of aggression were not so much to be found within individuals as within groups, especially when any of those groups considered itself to be a nation, a race, or a state.

## Group aggression

Luard (1986) considered that aggression at the individual level was largely irrelevant. Some 10,000 years ago, when settled food production began, there were clearly rewards for organized aggression. Groups would get richer and turn into empires if they were both aggressive and successful. Allport (1960) also moved his attention from individuals to group behaviour in his later report. Man's basic insecurity led to the formation of groups which soon developed mythological self-images. Groups could by psychologically conditioned for aggression by distorting images of neighbours or of the outside world and by creating an atmosphere in which violence was increasingly expected. Darwin's *Origin of the Species* has at various times been directly influential in providing an explanation or justification for aggression with its unconscious implication that it was inevitable that cohesive groups or societies with internal discipline and well-developed culture would expand at the expense of others (Darwin 1972).

The Darwinian line of argument is fiercest when linked to nationhood, race or statehood. Howard (1978) explained in *War and the Nation-State* how identification with the ideals of 'king and country' has been thought natural while international loyalties and subnational loyalties have had an uphill struggle. Klineberg (1974) argued that racial identity is the most motivating form of group identity, despite the fact that there is no other field where the conclusions reached by biological and social scientists differ so strongly from popular feelings. Race exists mainly in the minds of men, which in certain ways makes it more effective: it is flexible because it is not based on fact. For example, the Nazis were able to use racial self-enhancement when they were trying to win the German populace over to aggressive policies, yet they were also able to accommodate a positive image for the Japanese. Again, racial differences have been adapted to a number of colonial situations, for example to explain the Spaniards' destruction of supposedly inferior Indian culture or to portray empire as a 'White Man's Burden' without mentioning the economic gains to the imperial power. At another level, aggression in organized groups was encouraged by the whole range of racial stereotypes from Poles portrayed as idiotic to Hungarians portrayed as untrustworthy, from French letters to Dutch courage. Race and state are effectively combined in these kinds of example.

Assertiveness or aggression in individuals or groups is not enough *per se* to explain wars of assertion or aggression. Assertive or aggressive characteristics are best seen as necessary conditions for such war, but not as direct causes. So the possibility is there but not the explanation.

From aggression to war

There are a number of theories which seek to explain the link between assertion and aggression in individuals or groups and war, and they can be grouped into those which see such war as an outcome of liberal or romantic thought and those which see it as an instrument of political practice. It is the latter view which is favoured today, but it is worth first of all examining the former, which is firmly based in nineteenth-century thinking.

Great liberals like Mill justified the creation of empire for economic, cultural, and political reasons. Mill shied away from the inevitably violent consequences of obtaining territory for empires, but he set the scene for such violence (Mill 1975). The nineteenth-century liberal view, including Mill's, was that the Africans, Indians, and Irish were barbarous people who could reasonably be used to improve British economic circumstances so long as they were at the same time brought, however gradually, to the British cultural level. Freedom should only be given to states which could maintain their freedom, for example at that time Hungary but not Ireland.

Rosenblum (1978) argued that it was not liberal thought alone but its combination with Romanticism which could unwittingly turn political aggression into actual war. Romantic writers at the time of the French Revolution, the Napoleonic Wars, and the European Reaction are by no means always passionate about politics or inclined to be aggressive, but they are inclined to write about war as being the only way to enforce justice and as the very best method of self-expression. Romantics saw war as a form of action which overcame the limitations and frustrations of everyday life. Wordsworth saw war as the proper, spontaneous action of an inspired people, the beauty of the inner mind substantiated in an outward act, a glorious expression of a free people. De Musset saw war as a reaction against the anonymity of daily life. Byron saw death in war as a noble public act of self-expression and an opportunity for reabsorption into the unity of the universe. This was the kind of thinking that has often been admired and influential during the last 200 years. It was not eccentric, except in some of its expression, nor did it exist in a vacuum. It reflected a frame of mind which generated war among potentially aggressive groups of people, particularly if they were split into nations, states, and races (Cobban 1929). It was a frame of mind that never disappeared, although in the twentieth century it has been more usual to look for additional explanations for war and to see aggressive or assertive war mainly as an instrument of political practice.

Of course, war has never been divorced from political practice, but it is only relatively recently that war, along with taxation, economic development, and subsidies to farmers, has been seen as simply one of the alternatives available to government. The simple cynicism of Clausewitz was unfashionable up until the last fifty years. Clausewitz saw war as the inevitable direct outcome of politics governed by individual genius, chance, and irrationality (Clausewitz 1968). Twentieth-century thinkers like Aron considered it too simplistic to view war as merely a random outcome of political practice, or as the consequence of a romantic or liberal frame of mind. What was needed was a set of general causes of war.

Aron (1966) considered Hobbes's idea that the very survival of a given political unit was a general cause of war, but he found it inadequate. After all, man would not subordinate all his desires to his desire for life alone – no more would a state, a race, or any other group. There were, however, goals for which individual death would be risked; and the same was true for collective political units which wanted to be strong, to be feared, respected, admired, and capable of imposing their will on their neighbours and of influencing the fate of humanity and the future of civilization. Aron narrowed it down to desires for security, power, and glory, and found his examples in the history of France. Clemenceau sought security; his aim was to ensure that France avoided another war like the First World War. But an objective like this is ultimately unattainable because, in the end, no political unit can feel entirely secure until

19

it has no further enemies, until it is a universal state. Napoleon sought power – he wanted to rule Europe – but other states had to be constrained if they were to be commanded: power overwhelmed the calculation of interest. Louis XIV sought glory; his exploits were partly symbolic and he wanted to be glorious among monarchs. However, nobles who fought for prestige could never really stop fighting; like religious wars, the objectives of wars of glory are also ultimately unattainable.

In practice, security, power, and glory, as Aron defines them, are very difficult to separate. Unless he exterminates or drives out the inhabitants, the conqueror takes possession of both territory and the people who occupy it. Unless conversion takes place by the mere force of proselytism, the prophet does not disdain to govern before saving souls. Aron analysed the case of the French occupation of Algeria. The French sought security by eliminating the threat of Barbary Coast pirates; and they sought both power and glory, expanding France into Africa, 100 million Frenchmen in a greater France straddling the Mediterranean. On the other hand, the Crusaders sought only the power to liberate the Holy Land, not to convert Muslims. The sovereigns of medieval Europe collected provinces mainly because the glory of princes was measured by possessions. The superpowers seek security within their spheres of influence – Afghanistan and Czechoslovakia, Grenada and Nicaragua.

Identifying aggressive war

Every perpetrator of every war has sought to justify his adventure as moral and right. There is a substantial amount of literature on the morality of war. This literature can be used to help identify war which is purely assertive or aggressive, whether it arises directly from the philosophy and psychology of assertion or whether it is mixed up with liberal or romantic thought or with the demands of political practice seeking security, power, or glory. This kind of war will be a war of power through might, through the simple assertion of strength over weakness for its own internal philosophical and psychological reasons. Political units, normally states, which fight such wars may be morally resisted according to Walzer (1978), Johnson (1984), and others.

Walzer took the view that the British and French declarations of war in 1939 were moral because Germany was fighting an assertive war in precisely the way just described. In such a case, not only should the victim resist but, where the aggressor is tyrannical, there should be general war against the aggressor. Walzer expounded a theory of evil which overrides counting the number of lives involved in, for example, defeating Nazism.

Johnson criticized Walzer for his over-territorial approach to the way in which a war may be triggered. If it were moral that tyranny were to be opposed, surely territorial expansion would not be a pre-requisite for effective opposition. Additionally, terrorism and nuclear weapons open up whole new questions about the significance of territorial boundaries and the greater appropriateness of non-military forms of intervention like embargoes, diplomatic protests, and so on. Still, Walzer's recognition of assertive war is useful. He sorted out right from wrong not in an infallible way, but in a way which recognized the existence of the fallen world; that war can be a fulfilment simply or primarily of a desire for power and that war is often a calculated means to obtain it.

*Case-study 1.1: Nazi Germany*

Geopolitical models in Germany before Hitler

The practical consequences of the philosophy and psychology of aggression could be seen in Nazi Germany. Nineteenth-century German geopoliticians were the first to spell out these consequences, providing the whole theoretical basis of a programme for assertive war.

Foremost among German geopolitical thinkers was Ratzel (1969). For him, the nation-state was analogous to an organism, fluctuating in strength over the centuries and competing for space with other nation-states. It was space that lay at the centre of activity. All history was of world powers growing and declining in terms of space, in terms of territory. It was, of course, a close analogy with Darwin, but the true philosophical origin was the philosophy of aggression going back to Machiavelli. Ratzel argued that the ability to occupy organized space and the ability to assert authority were the crucial tests. In other words, the ability to win an assertive war was what mattered. It was inevitable and proper that successful states would expand their territory and therefore war was unavoidable. Ratzel never stated explicitly that war had to be military, indeed he talked of commercial and cultural competition, but he was a committed German patriot and his work was permeated with concern for German minorities around Europe. He made the point that a united Europe would have fewer wars, but he also made it clear that this united Europe would be dominated by Germany (Ratzel 1969).

By the 1920s the leading practitioner and founder of the *Arbeitsgemeinschaft für Geopolitik* (Centre for the Study of Geopolitics) was a romantic German nationalist, Haushofer. First, Haushofer accepted Ratzel's analogy between a nation-state and an organism as plain fact. Second, he was heavily influenced by Mackinder's view that the heartland of Europe – from the Urals to the Rhine – was the area which had to be dominated by any nation-state which sought power over the world-island of Eurasia and ultimately over the whole world. Third, Haushofer used geopolitics as a dynamic tool for making policy and staging political action. The result of all this was a mixture of eccentric scholarship with tendentious rhetoric and catch-phrases associated with oddly drawn maps designed to show how big Germany could be if all Germans were brought together into one nation-state (Fig. 1.1; Mackinder 1969; Parker 1985, ch. 5; Weigert 1942).

Haushofer was associated with the Nazis from their early days. His pupil, Hess, took him to see Hitler in prison, but he was never really committed to the primacy of Hitler's racial doctrine; for example, he never regarded Slavs as inferior to Germans, but rather as potential allies in ruling the heartland. In 1931, the Nazis effectively took over the *Arbeitsgemeinschaft für Geopolitik* and the fanatically Nazi Vorwinckel became increasingly important. All history, geography, and demography had to be reorganized under geopolitics and in accordance with Nazi racial doctrines. From Hitler's take-over in 1933, German geopolitics simply became the tool of the Nazi party, justifying whatever outrage happened next. Haushofer's personal influence went into decline, ending completely with Hess's flight to Britain and the German invasion of the Soviet Union in 1941.

### The geopolitical model of Hitlerism

The ultimate practical political consequence of the philosophy and psychology of assertion, expressed through German geopolitics, was Hitlerism. Hitler kept his ideology very much in mind when in power. His ideological refinement of power through might may be divided into his earlier doctrine of race and his later doctrine of territory. It was the latter which represented his main development of German geopolitics.

Hitler's starting-point was the defeat of Germany in the First World War, which he saw as a result of a conspiracy by Jews who, by the 1920s, he considered to be in control of Germany. The first stage of the required revolution would be when a 'nation-conscious' group took power. The second stage would be the destruction of Jewish power and also of decadent French influence. Hitler was therefore attracted by any political group dominated by the idea of a superior German people and he was influenced strongly by the anti-Semitic and anti-Slav pan-German movement based in Austria (Carr 1978).

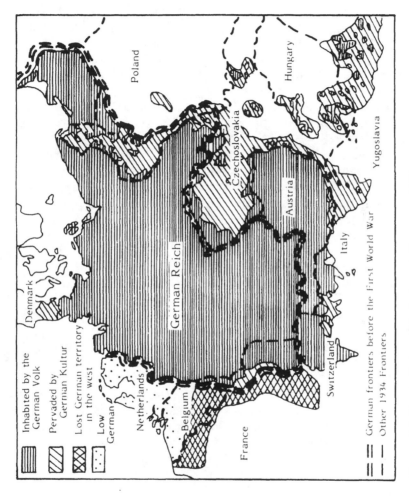

*Figure 1.1* Haushofer's political view of Germany, 1934

In prison, Hitler developed a more formal doctrine of race. The whole history of mankind was the rise and fall of civilizations which had either kept or lost their racial purity. It was a most extreme form of Darwinism applied to social science. The argument was that any restraint placed by civilization on human rapacity was a perversion of the evolutionary process which decrees that the fittest and strongest should survive at the expense of the unfit and weak. The biological urges to eat and to procreate are the foundations of all life, so when population outstrips resources, expansion at the expense of weaker neighbours is a right bestowed on the strong and a categorical imperative for the leaders of the strong. War, the inevitable consequence, is an unalterable law of life, the prerequisite for the natural selection of the strong. War is inevitable and desirable and brings about the triumph of Darwin's – and Fascism's – natural law.

Hitler's doctrine of territory was developed in terms of agriculturally usable land. His concept of a living space or *Lebensraum* was to do with having enough territory to grow food to feed a vigorous and healthy population. Indeed, Hitler sometimes interpreted the struggle between races as almost entirely a struggle for territory, just like Ratzel and other nineteenth-century geopoliticians, but with an open recognition that the implication was more or less continuous war. It was estimated that approximately an additional half-million square kilometres were needed immediately by Germany (compared with the 70,000 square kilometres lost at Versailles). Although diplomacy was also a recognized weapon, this additional territory would be worth an unspecified number of casualties. Cultural and trade weapons were not discounted, but the military option was always to the forefront (Hitler 1969).

This process of war through expansion and settlement by families who would in turn have further large healthy families and then demand further territory could never actually end, and Hitler recognized that it was an infinite process. The process could end ten or a thousand kilometres beyond the German borders: 'we National Socialists must hold unflinchingly to our aim in foreign policy, namely, to secure for the German people the land and soil to which they are entitled on this earth' (Hitler 1969: 596).

Hitler at various times made use of Ratzel's and Haushofer's arguments. For example, Hitler's assertion that the security of the state was directly related to the size of its territory came directly from Ratzel. But Haushofer's conclusions were in the end ignored by Hitler. Haushofer was only a conservative nationalist of the old school and would not preach aggression against the Soviet Union. On the contrary, in a world divided into space-owning imperialist

24

powers and space-deprived oppressed powers, he believed the Soviet Union to be Germany's obvious ally. In control of the Heartland (Eurasia), Germany in alliance with the Soviet Union could one day challenge the maritime powers and together become world powers. Hitler took from Haushofer only phrases and supplementary arguments.

Hitler's geopolitical viewpoint was quite specific and individual. The territory to be conquered was specifically to the east. In *Mein Kampf*, Hitler talked about the land between Germany and the Urals. The Slavs were regarded as inferior in every way and further weakened by the First World War and the Bolshevik Revolution. Complete falsification was used in his argument, for example the references to Slav lands as 'thinly settled' and the mendacious exploitation of the popular view of Poland as a '*Saisonstaat*', a nation-state for one season only (Hitler 1969: 586–609).

Hitler's practice of assertive war finally went wrong when he failed to dictate the war which he wanted to fight. He always said that he would put himself in the position of being able to wage war when it was appropriate, and he miscalculated the international reaction to Germany's invasion of Poland. First, he thought Britain and France would come round after the defeat of Poland and make peace, and later he thought that either lack of combat (in the 'Phoney War' period) or the heavy bombing of British cities would bring peace, but nothing worked. He thought that beating France would be sure to get Britain out of the war. When Britain had originally signed the Treaty of Assistance with the Poles on 25 August 1939, Hitler's first political instinct had in fact been to rescind the order to attack already issued. Why then did he go ahead in the end? Fest (1970) argues convincingly that Hitler, at this point, left the realm of intriguing politics which he had so far used to obtain his ends and instead followed his more natural instincts: his speeches became more doctrinal; he saw himself as a force of nature. It was the beginning of a dehumanizing process ending in the mass extermination in the Holocaust and uncalculated risks in the east.

## The fall of German geopolitics

Right at the end of the war, as old men and children fought in the ruins of Berlin, three dimensions of Hitlerism collapsed. First, the totalitarian involvement of every individual in a strict hierarchy with one man at the head (one of the main tenets of Fascism essential to the operation of Hitler's geopolitical model) became untenable. Second, the doctrine of race which endured to the last days of the war had important military consequences based on Hitler's geopolitical viewpoint. Third, the doctrine of territory brought about a military anticlimax in Berlin, but with a philosophical and emotional Armageddon for the defeated system of thought and behaviour (Slowe and Woods 1988).

The Nazi development of Fascism had also adopted the idea of natural law to free the hand of the government from any constitutional constraints. It had developed to its very limits the pivot of Fascist philosophy which was the involvement and intended self-fulfilment of individuals in the service of the state by using new propaganda techniques to ensure the total identification of obedient individuals with the one leader. This lasted right to the end of the war; it turned out to be a popular frame of mind which, once instilled, was very difficult to remove. Le Tissier (Tissier 1985) described the thoughts of a young German soldier in the last few weeks of the war who happened to keep a diary. The soldier, Helmuth Altner, newly arrived at the Front and terrified, was elated and entirely convinced of victory by the Führer's Order of the Day, and he expected a major success on Hitler's birthday which was a mystical date in the Nazi calendar; he found it mentally very hard to cope with the flight of the German army on that sacred day. This system of thought was also reflected in incredible acts of obedience and folly, which could have taken place only in a Fascist system where the will of the mass of individuals was subsumed in a 'general will' expressed by one leader; in fact it could only have taken place where Fascist philosophy was applied to a Darwinian concept of the world.

The doctrine of race was clear. The 'Master Race' had to fight the *Untermensch* from the east. Victory was inevitable; even if they were defeated now, the German people would rise again. There was also an overwhelming fear, fed by lurid propaganda, of sub-human hordes from the east. Hitler's doctrine of race was probably his most effective weapon at the end of the war because it created such a deep fear. Generals such as Heinrici and Busse sacrificed thousands of lives in their attempts to surrender as much as possible of Germany to the western allies and not to the Red Army. Entire battles were planned with only the direction of surrender in mind and the compulsion to escape west went through the population, to include the lowest ranks in the army and civilians alike (Slowe and Woods 1988: 74–82).

It was the failure of the doctrine of territory which at last brought Hitler's Germany to its end; Hitlerism was incapable of fighting a defensive war in a confined area like Berlin. It could only operate a system of settlement, enslavement, and expansion. As Berlin was surrounded, the collapse came quickly. Hitlerism needed population and territory to manipulate, it was a whole system of involvement, a whole set of hierarchies which needed lots of people and land to work properly. The will of the leader extended to everyday life in the Nazi world and it lacked the organization necessary for the purely military defence of a city. The last of the German army had to be thrown away fighting on the Oder in the east still dreaming of conquest, because there was no way of accommodating in Hitler's society his generals' practical plans for military urban defence (Slowe and Woods 1988: 37–40, 144–55).

Berlin fell quickly. Hitler committed suicide and the new Soviet administration quickly disposed of his body and destroyed his chancellery and bunker, leaving no shrine. When the assertion of power by a Fascist state failed, not just Fascism but, for a while, racism and the whole field of geopolitics on which Hitler's doctrine of territory had been based, were discredited. But it could only be for a while, for the Utopians and the Romantics and all the other components of the philosophy and psychology of assertion inevitably reappear in new forms at new times and in new places. Despite its worst defeat, the ability and willingness to use brute force is still a significant source of power.

*Case-study 1.2: Amazonas*

Geopolitics and expansion in Brazil

The Amazonas illustrate some of the practical consequences of the philosophy and psychology of assertion translated into action. This can be illustrated both by the expansion of the Brazilian state into the underdeveloped empty or Amerindian territories of its Amazon interior and by the conflicts between Ecuador and Peru as each claims the disputed Amazon Triangle. Here, Brazil is discussed first and then the Amazon Triangle.

The pioneer of Brazilian geopolitics was Backheuser. He was influenced by Ratzel and in his first geopolitical study of Brazil in 1925, he discussed Brazil's natural right to rule the interior of South America (Hepple 1986). Brazilian geopolitics matured with the publication by Travassos of a geopolitical look at Brazilian history, the *Projecao Continental do Brazil* (Travassos 1935). In this book Brazil's development was seen as progressing from the Atlantic coast inexorably inland, an organic growth. Travassos emphasized that for Brazil to develop this westward advance must continue towards Bolivia and up the Amazon. The movement westwards was seen as essential to provide living space for a growing and vigorous population. Travassos wrote, like Ratzel, of a revitalization of dead

frontiers into living frontiers; he believed this movement would lead
to the rise of a continental greatness ( *grandeza*) for Brazil. The
frontier was not a fixed line on a map, but a movable boundary
which stronger states could cross if they wished or needed. This was
especially true for regions which were seen to have great strategic
or economic value like the Amazonas. It led in turn to the 'Law of
Valuable Areas', which argued that if a nation failed to occupy and
exploit the resources of its whole territory then other states would
step in and take over (Child 1985: 107, 175). The path to *destino
e grandeza* (destiny and greatness) for Brazil was seen as inevitable
if she followed the geopolitical prescription of filling and exploiting
her empty or commercially unused interior spaces of the Amazonas.

In 1964, a Brazilian military government which was heavily influenced by Backheuser and Travassos came to power. It was different
from previous military governments in that it was not intended to be
just a short military intervention in civilian government but a regime
willing to accept responsibility for security and development. The
military came to power in response to a perceived threat from the
left, believing the only defences against such a threat were strict
security measures and a long-term economic development programme. Many of the basic concepts of Brazilian geopolitics,
including the organic vision of the Brazilian nation-state, could now
be translated into action.

The Brazilian military were directly influenced by an institution
which was permeated by geopolitical thought, the *Escola Superior de
Guerra*. This was a war college modelled on American and French
institutions, founded in Rio de Janeiro in 1949 by officers of the
Brazilian Expeditionary Force who had fought with the Allies in the
Second World War. Under the influence of geopoliticians like
Golbery de Couto e Silva and Meira Mattos, the *Escola* became
Brazil's first think-tank with its goals being

> to discuss and publish objective studies on theoretical and
> practical aspects of national security; to study and test the
> methodology for formulating and developing the policy for
> national security, including the relevant planning techniques; and
> to develop the practice of co-operation between sectors, thereby
> fostering a high degree of understanding between the individuals
> concerned, and promoting effective collaboration between the
> various sectors responsible for national security.
>
> (Kelly 1984: 442)

National security was the theme, and national security in Brazil
entailed a practical response to geopolitical ideas.

There were two types of courses taught at the *Escola Superior de*

*Guerra*, the graduate war course for officers and civilians and the armed forces general staff course for the military only. By the 1980s nearly all active-duty generals and over half of the colonels in the army had graduated from the *Escola*, with a number reaching high posts in the government, especially in the president's office and the national intelligence service (Stepan 1971).

The most important geopolitical idea promoted by the *Escola* in the 1950s and 1960s originated with Golbery de Couto e Silva. General Golbery was the first Brazilian geopolitician to apply geopolitics directly to government policies. He was able to do this not only because of his influence over the *Escola*, but also because he was an important presidential adviser for nearly fifteen years after the 1964 coup. *Aspectos Geopoliticos do Brazil* set out Golbery's view of the application of geopolitics to domestic power politics (Golbery 1957). Along with his acolytes at the *Escola*, he developed the concept of 'the permanent national objectives of security and development'. Golbery saw Brazil as beset by enormous problems, subject to left-wing subversion, with the only way of protecting society being to industrialize rapidly and, at the same time, to apply strong internal security measures: these were to be the permanent and paramount national objectives. In 1967, Golbery said that to struggle to survive requires maximizing economic growth. Brazil could either become great or perish. This was the basis for the doctrine of 'security and development', an approach to political problem-solving with the emphasis on economic development, particularly industrialization, as a source of national political power. At the same time, the role of national security would be to save the nation from itself; and this could include the neglect of individual human rights in favour of the interests of the community, as perceived by the ruling military.

The military objectives, the government's refinement of Golbery's 'permanent national objectives', were

> to establish a national community, politically, socially,
> economically, and culturally integrated; to guarantee the exercise
> of complete national independence; to maintain territorial
> integrity; and to project the national personality in the concert of
> nations.
>
> (Selcher 1977: 12)

Meira Mattos, another instructor at the *Escola*, was also a major influence in the early 1970s on the way in which the 'permanent national objectives' were put into practice. Mattos expanded on Backheuser's and Travassos's concepts of organic frontiers and the liability of apparently vacant interior spaces. He advocated enhancing territorial consolidation without any aggression towards neighbouring states like Argentina but without regard to any indigenous population. Mattos, like Golbery, believed that only through economic development and expansion, and the political power which these would bring, could the security and progress of Brazil be assured and her *grandeza* achieved. Mattos repeated Ratzel's dictum that space is power. He even took the idea a stage further by saying that territory conditions the whole life of a state and limits its appreciations. He saw Brazil as being in the right location to tame the frontiers of the Amazon, with the great advantage that the frontiers for Brazil were within its state boundaries, therefore its expansion would not need external aggression.

> It would be dangerous to leave the vast Amazon Basin empty and underdeveloped when there are areas of growing overpopulation such as Bangladesh and Indo-China. It would not be desirable to the Amazon countries to lose their sovereignty over this coveted region under the pretext of their incapacity to exploit it.
>
> (Kelly 1984: 444)

Just as Mackinder had seen Eurasia as the heartland and the pivot of the world, Mattos saw the Amazonas as the heartland of South America.

Brazil's development of European geopolitical ideas was clearly visible in the policies of the military governments, especially those concerning the expanding Brazilian internal frontier. A clear frontier policy came under the general programme of spatial and national integration. The military governments' policies sought to minimize regional diversities, occupy perceived empty spaces, and increase the rate of discovery of natural resources and the accessibility of those resources for exploitation. Their efforts went into improving transport and communications and encouraging settlement programmes. The emphasis was to be an advance into the Amazonas heartland at a rate which would not be seen as aggressive by neighbouring states.

New laws were passed specifically to promote development in the Amazon. These were aimed at creating development poles and self-sustaining population groups in the frontier regions. Great emphasis was placed on a rapid industrialization programme to be financed by foreign and domestic corporations. The occupation of the region was mainly to come from internal migration. This was all precisely in line with Brazilian geopolitical ideas, just at a time when Travassos's 'Law of Valuable Areas' was being enacted in neighbouring Amazonian states, so it also dealt with a perceived military threat from Peru and Venezuela as they embarked on similar development programmes of their own. A special government office for the development of Amazonia was set up to help in the speedy organization of people and companies wishing to move into the Amazon area. A variety of financial incentives was added, and Manaus, the largest city in central Amazon, was made into a Free Trade Zone. In March 1967, a five-year plan for the Amazon was adopted; it was expensive, ambitious and with the overriding problem of vast dependency on government finance and foreign finance (Wesson and Fleischer 1983).

Inevitable problems arose. New programmes were developed and revised under subsequent civilian governments. Brazil's expansion into its living frontier was and is expensive and often destructive. It still continues apace. It is power through might in action, perhaps fortuitously within a single state.

Geopolitics, expansion, and war in the Amazon Triangle

There are many similarities between the Brazilian claim to the Amazonas and the claims made by Ecuador and Peru. The main

*Figure 1.2* The Amazon Triangle

difference is that while Brazil's claim to the main Amazon Basin is internationally undisputed, Ecuador and Peru both claim the same area: the Amazon Triangle bounded by the Marañón River, the Peru–Colombia boundary and the disputed Peru–Ecuador boundary. Both Ecuador and Peru, like Brazil, take an organic view of the state, based originally on European geopolitics. Both also share Brazil's view of the 'Law of Valuable Areas', that areas which are empty of people except Amerindians, but full of possible mineral and forest resources, should be occupied or they are likely to be lost to rival states.

Peru sees herself as a small state surrounded by threatening neighbours. Her military men, like their counterparts in Brazil, have a keen sense of geopolitics, illustrated by this quote from an interview with a senior Peruvian general:

> We Peruvians must buy many weapons because, like Israel, we are surrounded by enemies: Chile to the south wants to refight the War of the Pacific [in which Chile had won substantial and valuable territory from Peru and Bolivia in 1879]; Ecuador to the north wants to steal our Amazon territory and our oilfields; Colombia to the north-east has not forgotten the 1932 Leticia incident [when Peruvian citizens crossed into Colombian territory and invaded the town of Leticia in support of Venezuela in a minor territorial claim against Colombia]; and there is Brazil – Brazil which, like the United States of a hundred years ago, believes she has a manifest destiny to occupy the continent and reach the Pacific. And in South America, Peru is California.
>
> (Child 1985: 98)

Ecuador takes a grandiose view of the state it ought to be. Geopolitical thought there promotes a 'Greater Ecuador' which includes the Amazon Triangle finally lost to Peru in a war in 1941. José García Negrete, an Ecuadorian political geographer, sees Ecuador as being like the god Janus with two faces, looking at the same time inland to the Amazon Triangle and outwards to the Pacific Ocean where Ecuador reaches the Galapagos Islands. The 1942 Rio de Janeiro Protocol, which followed the war with Peru, is regarded as an unequal treaty which has mutilated Ecuador and deprived it of its greatness; the Ecuadorians call it the 'Protocol of Sacrifice' (Negrete 1972).

The legal origin of the dispute over the Amazon Triangle lies with the ambiguities and uncertainties of the Spanish colonial division between the Audencia de Quito (Ecuador) and the Viceroyalty of Peru. The so-called North Andean dispute simmered through the nineteenth century and the first decades of the twentieth century with both sides presenting legal arguments and Peru exercising actual control over most of the contested area. In the late 1930s, a series of boundary skirmishes led to attempts by outside arbiters (chiefly Argentina, Brazil, and the USA) to reach a peaceful settlement, but Peru discouraged these efforts. Peru was basically eager for a fight which it could win. Her major economic and territorial war with Chile was lost and it was said that the army needed a tonic! (Child 1985: 92–8).

Peru regards the Ecuador boundary as settled, except that Ecuadorians like periodically to stir up trouble. The main theme of current Peruvian geopolitical thinking towards the Amazon Triangle is the need to develop it economically as part of a policy of 'integral security'. Economic development of the Triangle, which may include exploiting uneconomic oil resources for geopolitical reasons, helps to secure the area's links with the mainstream Peruvian economy and Peruvian society. The border areas themselves have already received substantial investment compared with other remoter parts of Peru; for example an expensive jungle road has been built along the disputed boundary, the *Carretera Marginal de la Selva*.

Ecuador accepts none of this. For Ecuador, the 1942 Protocol was unequal, unfair, and forced on a state at a time of weakness when foreign forces occupied a good deal of her territory. Peru occupied substantial tracts of Ecuador by the end of the 1941 war. The same war flared up again in 1981. There are no signs of a permanent settlement.

Conclusion

The Ecuador–Peru conflict is not just a boundary dispute. It is possible to distil a legal argument about the boundary from it but that is not at its heart. Both Ecuador and Peru have clear geopolitical self-images. Like Brazil, the conquest and economic development of the Amazonas, an inland frontier, are a vital part of that self-image. The conflict is geopolitical.

In Brazil, Ecuador, and Peru, there is a direct, traceable link between geopolitical thought and geopolitical action. A desire for the assertion of the state over territory which is seen as virgin territory is followed by geopolitical theory to justify whatever action is necessary to secure that territory. Then the people who really believe these things take power in the state. They either are themselves the military or are backed by the military eager to flex its muscles. It is power through might – in action.

*Case-study 1.3: South Africa and Namibia*

The origins of white domination

The long struggle for control of South Africa between the English and the Afrikaners (originally Dutch) ended in 1948 with independence for white South Africa, finally abandoned to the Afrikaners who formed the majority of the white population. From independence onwards, much of the zeal that had gone into retaining Afrikaner identity under British rule was channelled into retaining South African white identity as a whole. The Afrikaner concept of apartheid, which saw whites in general as being racially superior to blacks, was in straight succession to the Calvinist concept of the Afrikaner as the elect of God. Although there had always been a South African political claim, used especially in international circles, that apartheid seeks nothing more than separate development with no superiority and no inferiority, in fact apartheid is based precisely on a philosophy of superiority and inferiority, backed up by the imposition through brute force of white over black economic and cultural power. Apartheid in South Africa has been an example in theory and practice of the assertion of power through might.

Giliomee (1979) proves with many examples that apartheid is about identifying superior and inferior races. According to Afrikaner ideology, mixed marriage is evil because it degenerates the white race; whites have to be the only ones to vote because they were the founders of the prosperity of the South African land; the difference between black and white is the physical manifestation of the contrast between two irreconcilable ways of life, between superior and inferior, between barbarism and civilization, between heathens and Christians, between overwhelming numbers and a few defenders of the high and holy ground.

Superiority and the sense of being swamped by larger numbers of blacks made instant nonsense of the idea of apartheid based on whites and blacks arriving simultaneously in an empty land and both being entitled to their own areas for equal separate development, which is the official line. The feeling of most Afrikaners was that if whites were superior they were entitled to dominate and if there were fewer of them they had to dominate to survive.

The outcome of this feeling of superiority and this need to dominate is the assertion of white power in South Africa today. Its origins are to be found in the South African Reformed Church's interpretation of Calvinism and the economic self-interest of South African whites. These together have combined to create the Afrikaner idea of apartheid and to encourage the use by South African whites of armed might to dominate South Africa and Namibia.

34

The South African Reformed Church bases its teaching on the central concept of Calvinism theology: predestination. God chooses certain people for salvation in advance of their birth. There is no sure way of knowing who is among the elect, but indications include leading a religious home life and a diligent and successful commercial and community life. This tends to be accompanied by an attitude of exclusiveness, where the attitude towards those who do not show the right signs is full of hatred and contempt because they must be a condemned enemy of God. In other words, Calvinists, confident of their own grace, are contemptuous of others who must struggle almost certainly in vain. With this harsh creed in mind, it is easy to understand South African believers' contempt for the black non-believers surrounding them, and the majority of Afrikaners are such believers, predominantly members of the main white South African Reformed Church, the *Nederduits-Gereformeerde Kerk*.

When the General Synod of the Kerk met in 1966, the doctrines of apartheid were agreed. The Kerk sought to clarify the theological explanation for white racial superiority and at the same time to explain the central importance of the assertion of this superiority in South Africa.

First, the Synod confirmed the importance of predestination and discussed purgatory as a punishment for the predestined who have sinned and the impossibility of reward for those who are good but lack faith.

Second, the Synod declared that ethnic diversity was in accordance with God's will and that this resulted in the formation of many different races. When man tried to unite as one race it was against God's will; after all, God had destroyed the Tower of Babel. God gave every race its own distinctive characteristics and these gifts are put at risk by racial mixing. Thus, blacks who become Christians should not be integrated with the white community, for integration would be against God's will, but should be told to show other blacks the faith.

Third, God gave each race its own land: 'When the Most High parcelled out the nations, when he dispersed all mankind, he laid down the boundaries of every people according to the number of the sons of God' (Deuteronomy 32:8). As a result of sin, boundaries and races had become all mixed up, but a small number of righteous white men had come to the Cape where they had formed a community loyal to their faith. When British commercial interests threatened their exclusive ways, they made the 'Great Trek' inland and settled in the north of their chosen land, which they then defended against heathen Matebele and many others. They met with hardship, they fought against the British, and finally they won back their land in 1948. It would always be threatened by the heathen hordes and would always have to be defended. The Voortrekkers who had made the Great Trek of 1838 had demonstrated their state of grace by quitting the British commercial ports (just as Abram had chosen the hardship of the Chosen Land in preference to Sodom and Gomorrah) and had carved out their own new land and saved it from barbarism and neglect. Now it fell to their descendants to prove themselves worthy of their forebears by leading good and prosperous lives and to defend white South Africa (General Synod of the Dutch Reformed Church 1966: 5–8).

### The contradiction between white theology and economics

There was a direct contradiction between the Kerk's instructions on the separation of races and the need to prosper and defend Afrikaner land. The nature of economic growth, which fulfilled the injunction to prosper, had to contravene the injunction to separate the races. The non-white labour force outnumbers the whites in all South Africa's major cities by two or more to one. Black labour underpins the whole of the South African economy. In the last ten years, shortages of skills have forced whites to recruit blacks even into supervisory and professional jobs. The theological demands of apartheid theory cannot work in an industrial society dependent on black labour. Instead, apartheid is used merely as an excuse to oppress black labour or as a justification for the South African government policy of the moment, just as Hitler referred to Haushofer's geopolitics when it was convenient. Like the concept of greater Germany to Germans, the idea of apartheid to South Africans is attractive to the self-image, but it is not enough on its own to explain the assertion of white South African power. There is the vital added ingredient of South Africa's role in the international economy.

South Africa plays a part in the periphery of the international economy. The development of capital in the periphery is blocked by the greater competitive strength of industries in the core developed states. The periphery's lower wages cannot generally compete with the core's higher productivity, so the main source of income for the periphery becomes primary goods over which the core states have control of prices. The bulk of the South African economy is peripheral in this way: foreign trade is a large fraction of the gross national product, there is a heavy dependency on imported capital goods and on the earnings of exported primary goods. In short, South Africa's trade patterns are those of a Third World state. There is also typically an elite which holds political and economic power far in excess of most of the population.

There are, however, certain major differences between South

Africa and the rest of the Third World, of which the first and most obvious is that apartheid had dictated that the elite is white whereas the rest of the population is black. There are also two further important differences between the South African elite and other Third World elites: first, there is the large size and high life-style expectations of many of the white South African elite; second, there are the problems arising from the international repudiation of the South African system.

The white elite of South Africa is large and requires a life-style similar to Western Europeans and Americans. The whites of South Africa do not see themselves as a small Third World elite at all. They see themselves on equal terms culturally and commercially with white western states. Many of the skilled English-speaking whites attracted to South Africa in the last thirty years went there on the understanding that they were going to a place which could provide them with more comforts and a better standard of living than Western Europe. To secure and maintain such prosperity for 30 per cent of the population, a much greater elite than in any other Third World state, requires a degree of exploitation of the rest of the population far more organized and effective than would normally be necessary. To meet such a requirement, apartheid theory provides the perfect ideology: the self-righteous subjugation of 70 per cent of the population can still just provide South African whites with prosperity comparable with Western Europe.

Then there is the international repudiation of apartheid. The direct economic impact of this repudiation has been minor, but its political effects have forced South African whites to devote substantial resources to defence and security. This has put pressure on the economy to expand in fields where it can make quick export earnings. Just as in South America, where the Law of Valuable Areas dictated the economic exploitation of remoter parts of the state to prevent their loss, so South Africa intensified its exploitation of the territory it controlled to the north, Namibia. Like the Amazon Triangle, Namibia was contested territory, but it was for the time being under South African control and offered the chance of rapid and profitable exploitation. Its rich mineral resources easily attracted the necessary foreign capital, and massive dollar earnings soon followed.

The theology of apartheid provided the key to the Afrikaner self-image on which modern South Africa was built. The Afrikaners were the elect of God. Theirs was the right to dominate the heathen and theirs was the duty to keep apart from other races; theirs was the right to be a white elite. Their participation in the international economy put them in a dilemma. They could not maintain their

37

position without involving blacks in their economy and therefore mixing with them. They found they had to share their land but at least they could keep barriers within it, the townships were separated from the cities, the black homelands from the white heartlands in the bizarre geography of apartheid. At the same time, as part of the same economic development, the Afrikaner self-image became blurred with the general South African white self-image. As economic considerations in the second half of the twentieth century came to take precedence over theological considerations, so the preservation of the position of the white elite became more important than the preservation of the exclusiveness of the Calvinist elect. What was ideally wanted was an opportunity for white economic gain without further racial compromise with blacks and without increasing the size of the white elite and thus risking lower living standards; Namibia presented just such an opportunity.

The occupation and exploitation of Namibia

Namibia, or South West Africa, had been a German colony until 1915, when it was given up after the First World War. Article 22 of the Covenant of the League of Nations provided for:

> those colonies and territories which, as a consequence of the late war, have ceased to be under the sovereignty of the states which formerly governed them and which are inhabited by peoples not yet able to stand by themselves under the strenuous conditions of the modern world.
>
> (Laurus 1965: 372)

Namibia was handed to South Africa which was then part of the British Empire.

The ranching and fishing industries of Namibia, and limited amounts of mining, developed under largely disinterested British South African rule in the interwar years and up until the early 1960s. It was only then, in the 1960s and 1970s, under the South African Prime Ministers Verwoed and Vorster, that the twin domestic economic and international political pressures really took effect. Additionally, the first serious local and international pressure for Namibian independence was applied. For all these reasons, an expensive and ambitious programme was launched to enlarge and modernize South African armed forces and a period of intensive South African involvement in Namibia began.

Until then, South Africa had been content to support the resident German whites and a few South African and foreign companies to exploit what they wanted in Namibia. All this changed dramatically in a few years. A whole series of mines opened up, exploiting with astonishing speed Namibia's rich resources of copper, lead, zinc uranium, diamonds, vanadium, silver, tungsten, and tin. The GNP per head in 1960 in Namibia was £75 and the average wage for a black Namibian was £55; by 1983 the figures were £760 and £110 respectively (International Labour Office 1977).

The maintenance of the necessary high incomes for South Africa's skilled European expatriates, and for South African whites in general, has been possible only with the kind of colonial exploitation that has taken place in Namibia. The colony cannot remain forever. South Africa's expensive involvement in Angola to try to defeat the radical Namibian independence movement (the South West African People's Organization) has drained some of the profits, but the rapid exploitation of Namibian resources carried out in the light of the possibly short period they would be available to South Africa has made the profits enormous (Katjavivi 1988). The involvement in Namibia has achieved its immediate purpose of enhancing the living standards of the South African white elite.

South Africa looks like trading Namibia for the withdrawal of Cuban troops from Angola and the ending of direct Soviet involvement in the area (*The Economist* 1988). For a time at least, South Africa should then be able to achieve in Namibia by economic dominance the same as it has been achieving by armed force. South Africa has the accumulated wealth and accessible resources to dominate economically its politically unstable and economically weak black neighbours. It is likely for some time to be in a position to put into practice by economic means its particular version of the philosophy of power through might.

*Case-study 1.4: The Western Front in the First World War*

Disputes reduced to tests of power through might

Once war has been declared, the only way to victory is by brute force. The success of diplomacy, trade embargoes, blockades, and all other means more peaceful than battle depend on the battle itself. Whatever the origins or causes of a dispute, the battlefield becomes the practical testing ground and armed force the basis of resolution. Military geography takes over; territory is fought for, won, and lost in a pure test of strength and battle skills.

One example of a complex of political and economic problems reduced to territorial battle, is the First World War. Slowe and Woods (1986) in their study of the Western Front in Flanders and Northern France provide a number of examples of different ways in which territory was won and lost in battle in the First World War.

Two main simple methods predominated as they always have: weight of numbers and surprise superiority of weapons.

Methods of gaining territory

Probably the best-known case of the conscious reliance on sheer weight of numbers to win territory in battle, certainly in the First World War if not in all history, was the battle for Passchendaele, the Third Battle of Ypres. Throughout the height of this battle a vivid diary was kept by Captain Edwin Campion-Vaughan which illustrates the relentless push forward to capture territory which had no importance in itself but which came to represent victory or defeat for Britain or Germany (Figure 1.3).

Campion-Vaughan's company took part in the attack by waves of British infantry on Langemarck Ridge, a range of low hills 4 miles east of Ypres. The first wave had started up the ridge several hours ahead of Vaughan, and most of them had been killed by the German machine-gunners defending the ridge or had drowned in the mud by the time Campion-Vaughan himself set off. The conditions were atrocious from the flooded Steenbeck Valley onwards:

> I paused for a moment in the shell-hole; in a few seconds I felt
> myself sinking, and struggle as I might I was sucked down until I
> was firmly gripped round the waist and still being dragged in.
> The leg of a corpse was sticking out of the side, and frantically I
> grabbed it. It wrenched off and casting it down I pulled in a
> couple of rifles and yelled for the troops in the gun pit to throw
> me more. Laying them flat I wriggled over them and dropped,
> half dead, into the wrecked gun position.
>
> (Quoted in Slowe and Woods 1986: 56)

Campion-Vaughan led the second wave of attackers up the ridge. Many more were killed outright or were wounded and left to drown. Eventually he reached the top and his company captured Springfield Farm which had been converted into a machine-gun emplacement. It had been a day's journey from the valley, 500 yards horizontally and less than 100 feet up to the top of the ridge. In the evening, Vaughan's company were relieved by yet another wave of infantry who were to attack new German positions beyond the ridge. The Germans were forced back by a British willingness to take losses as the simplest way to advance in the battle (Figure 1.4).

Occasionally, a similar territorial result could be obtained by a surprise superiority of weapons with far less human cost to the attacking side. This was the case, for example, when the Germans first used flame-throwers, when a good deal of tactically important territory fell in a few minutes. But the most spectacular case on the

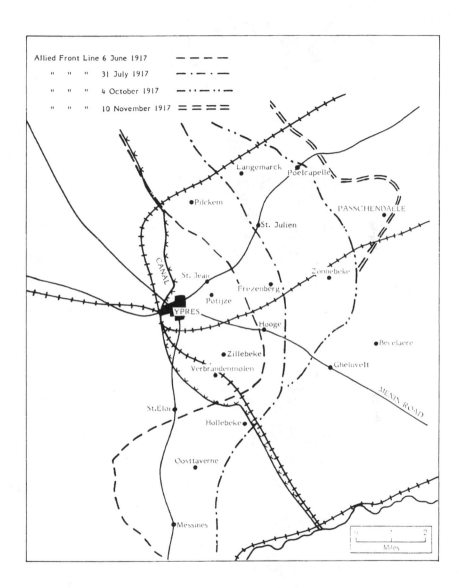

Allied Front Line 6 June 1917 — — — —
"    "    "    31 July 1917 — · — · —
"    "    "    4 October 1917 — · · — · · —
"    "    "    10 November 1917 ═ ═ ═ ═

*Figure 1.3* Ypres, 1917

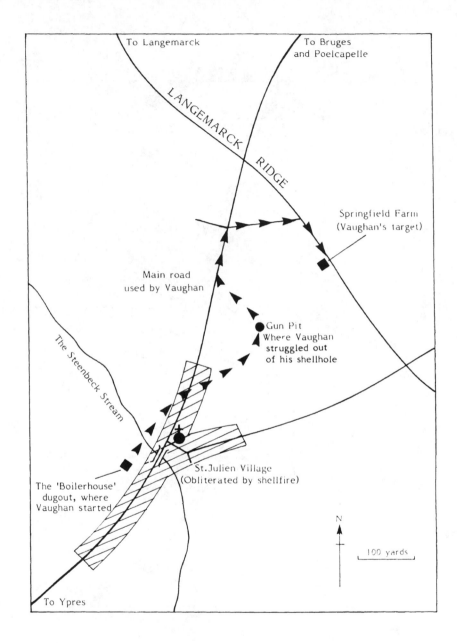

*Figure 1.4* The route taken by Captain Edwin Campion-Vaughan in the Third Battle of Ypres

Western Front was the first use of the tank, originally in ones and twos on the Somme, but later in large numbers, as in the Battle of Cambrai:

> The tanks rumbled in their hundreds over the top of the slight slope which marked the German front line. There was no effective resistance to them. Soon the artillery stopped firing . . . and it was time for the infantry advance. It was one of the easiest infantry advances of the war. In four miles, [Corporal George] Coppard [of the Machine-Gun Corps] was stopped only once . . . a German machine-gun had been hidden by the road embankments from the first line of tanks and was taking its toll of British infantry. Seeing what was happening, Coppard managed to attract a second-line tank crew's attention. The tank lumbered to the sunken road and turned up it. The crew spotted the machine-gun post, and the machine-gunners with the terror of death realised the tank was coming for them. They fired in panic. The bullets bounced off the steel of the tank. The tank's light cannon ponderously and unstoppably turned on the machine-gunners and killed them. Then the tank crushed them. The way was clear.

But there was a catch:

> At first Coppard though the tanks had won a massive victory, wiping out all the Germans. Then it dawned on him that there were only a few dead Germans and no destroyed artillery pieces. Of course, it would have been impossible for 400 tanks and an army of artillery and infantry to be gathered without the Germans knowing anything of what was happening. The great tank battle of Cambrai had been an important and exhilarating advance, . . . but it was not a hard-won victory on the day.
>
> (Slowe and Woods 1986: 207–9)

Although it was possible to win territory by other methods, especially by a surprise superiority of weapons combined with appropriate tactics, it was practically impossible to win a really significant battle without the kind of bloodletting that went on at Langemarck Ridge. It was just the same in the Second World War, where no new weapons could have substituted for the inch-by-inch fighting either at Stalingrad or on the Pacific Islands, and it was just the same in the Iran–Iraq war, where no amount of poison gas could substitute for the hand-to-hand fighting in the marshes east of Basra. Military geography demonstrates the simple truth that at the level of the battlefield it is mainly through human sacrifice that one side can assert its strength over the other.

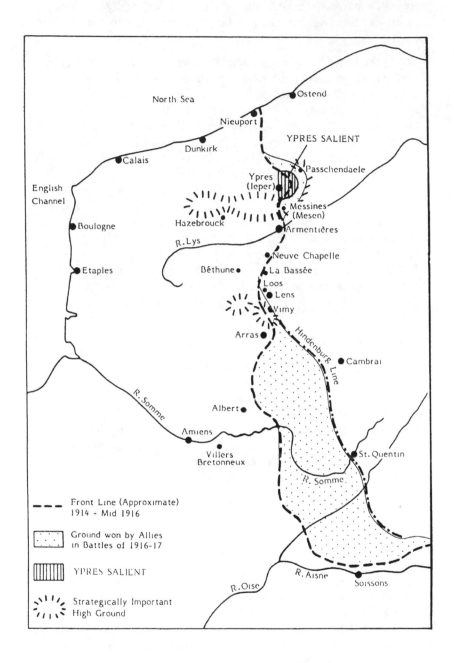

*Figure 1.5* The Western Front 1914–18, British Theatre of Operations

Battlefield psychology: enjoying the fight

What makes these battles possible to fight effectively is the naturally aggressive nature of many of the people who get caught up in them. For many of the individuals involved in these battles, it was not all horror as it might seem to the reader afterwards. While it would be far-fetched to say that many of the participants actually enjoyed fighting, it was certainly a highlight of most of their lives. Steiner (1971) saw the war in general terms as the outcome of a long period of boredom and frustration in industrial society. Slowe and Woods show what battle could offer people for whom life was drudgery and boredom:

> The men in the front line experienced the very extremes of the emotions of fear, comradeship, hate, pity – far beyond nearly every experience of ordinary life. It is these emotional extremes which are the very essence of life in battle.
>
> (Slowe and Woods 1986: 231–2)

At the time of the outbreak of the First World War, life for most young men was particularly unattractive and dull. Long hours at monotonous work for low pay were typical. So for many, the war offered a way out to adventure and perhaps glory too. Most soldiers were not casualties; many had great stories to tell when they got home. For them the war created a sense of belonging and purpose, despite all the tedium and discipline of the day-to-day life of a soldier and the mental and physical scars some carried for the rest of their lives. This was a new experience, an escape from routine and a chance to participate in a victory.

Battlefield psychology: memorabilia

The conditions in battle were often so ghastly that words can scarcely be found to describe them; yet, afterwards, battlefields were turned into victory memorials and the stories of battle into *Boy's Own* tales of heroes and villains. The battlefield became a symbol of the adventure, comradeship, and sense of purpose of physical fighting. For the men who benefited from the war in self-esteem and self-fulfilment and for the states which might once again need to resort to battle, this was a useful and positive symbolism. It is a good final example of the philosophy and psychology of power asserted through armed might to see the way in which the battlefields of the First World War became the shrines of heroes in the years that followed. Continuing the theme of the First World War, a fine example would be the Ypres Salient (Figure 1.5).

The First Battle of Ypres was in October and November 1914, a series of skirmishes compared with later battles but still important enough to wipe out a fair proportion of the British regular army's Expeditionary Force. The Second Battle of Ypres, from April to July 1915, saw the first use of chlorine gas by the Germans in their attacks, and casualties mounted in the British Volunteer Army. The Third Battle of Ypres began on 31 July 1917 and continued until November; this was the battle known as Passchendaele, after the village which was its main objective. One-quarter of a million British soldiers, including many conscripts, were casualties. The Fourth Battle of Ypres in 1918 was part of the wider Lys offensive which helped to bring the First World War to an end. The retention of Ypres Salient through the war became a matter of national pride, involving difficult odds. It was to be defended to the last, and the British soldier would fight or die gloriously for it. The coming of peace brought changes of emphasis but much of this symbolism was as relevant after the war as during it, and an influential organization was set up entirely devoted to sustaining it.

The largest post-war association of ex-servicemen and others wishing to retain an interest in the First World War was the Ypres League which had a fortnightly publication, the *Ypres Times*. This mass-membership organization and its publication gained in stature and readership throughout the 1920s, and its influence in establishment circles went far beyond the members and readers. It became the instrument for turning Ypres into a heroic symbolic place and an announcement was made on Ypres Day (31 October) 1925 that:

> the name of Ypres will never be forgotten. Ypres symbolizes
> something which it must be the wish of all who care anything
> about the abiding things in life shall be preserved. Ypres
> symbolizes a spirit and a force – the spirit of devotion to duty,
> discipline, King, Empire, the spirit which puts the cause above
> the worker for the cause, the spirit which has always
> characterized England at her best. And it symbolizes a force –
> the force of endurance in the face of tremendous odds, of patient
> bearing of weariness and discomfort, of what we in popular
> speech call 'grit'.
>
> (Storr 1925: 4)

The romance of war played an important part in turning Ypres into a symbolic place. There was a constant evocation of the 'spirit' of Ypres, especially when justifying the work and existence of the League itself. People who were against the ideals of the League might have their views changed . . .

> by the spiritual ideal of men and women in its ranks who belong
> to it, not forgetting the tragedy of the war, not for morbid

motives or for militaristic aims but to keep in living
remembrance the splendour of the spirit of our people at that
time, their defence of national life and liberty, their exalted
resistance to all the powers of evil and darkness in league against
the nation and the permanent memorial of all that was noble in
the War.

(Gibbs and Waggett 1922: 177)

To develop this romantic evocation of battle spirit, a junior division
of the Ypres League was also formed.

May each new generation, as it comes along, remember what
Ypres stands for and, remembering, feel in itself the will to stick
up for the Right and also the Power to bring it to pass.

(*Ypres Times* 1924: 73)

Religion also played an important part in turning Ypres into a
symbolic place.

Ypres, at all events is a shrine, a holy place, to at least a million
people of the Empire who were never there.

(*Ypres Times* 1921b: 53)

The frequent use of the words 'pilgrim', 'pilgrimage', and 'shrine'
attributes a religious or holy status to Ypres. The battlefields of the
Ypres Salient were called 'Holy Ground':

As in the old days, our folk went in for journeys as pilgrims to
holy places to get refreshment of soul, more courage for life and
death, some closer spiritual touch, in their simply human way,
with the divine meaning of things, so now Ypres and those
cemeteries in the Salient are Holy Ground to which the spirit of
our race must go on pilgrimages.

(Gibbs 1922: 199)

Careful attention was paid to numerous memorials erected in
memory of the battles and the soldiers who died in them. For example, the official government war memorial at Ypres was the 'Arch
of Triumph' erected at the Menin Gate in Ypres. It honoured the
missing by inscribing the names of 58,000 officers and men missing
in action. The Arch fulfilled a special role every year as part of the
Remembrance Day service and the Last Post was played there every
night (and still is).

Finally, royalty also played its part in the process of making Ypres
a symbolic place. The King was patron while the Prince of Wales,
'a soldier who knew the Salient well' (*Ypres Times* 1921a), was vice-
patron of the Ypres League. Messages from the King and other

members of the royal family were often included, for example from the King himself:

> The Ypres Salient did not give way. People said they ought to withdraw, but behind the whole thing there was sentiment and the commanders in the Salient said: 'The Salient has been pierced but we will hold on to the bitter end'.
>
> (*Ypres Times* 1923: 18)

The constant use of, and reverence for, the name of Ypres, the romance, religion, and royalty were all part of a systematic promotion of Ypres (especially in the ten years after the end of the war) as a symbolic place.

Conclusion

Whatever the causes and purposes of the war of which the battles fought in the Ypres Salient were part, those causes and purposes were translated into tests of physical strength on the battlefield. These tests were usually resolved by the weight of numbers prepared to die, but superiority of weapons or subtlety of tactics played a part as well. The fighting around Ypres was the expression of power through might at its most pure and brutal. It is important to realize that such fighting was anticipated with relish by many a young soldier; they were keen to go to Ypres to see what it was like and to take part in the real thing. Afterwards, for many years, it was remembered by some with tears of sadness but by most with tears of nostalgia and pride.

**Power through might: summary table**

*Philosophy of power through might*

| | | |
|---|---|---|
| 1. | THE DOMINANT RULER OR THE DOMINANT ELITE | Machiavelli – the individual ruler in a territorial state. Enlightened absolutism – moderating the individual ruler or the ruling elite. Utopianism – liberating the individual ruler to do as he likes. |
| 2. | THE DOMINANT STATE | Nietzsche and Treitschke – the state as an organism. Fascism – the dynamic state. |

*Psychology of power through might*

| | | |
|---|---|---|
| 1. | INDIVIDUALS | Is man aggressive? |
| 2. | GROUPS | The role of nations, races, and states. |
| 3. | FROM ASSERTION TO WAR | The need for glory, power, and security. |

Case-studies relating GEOGRAPHY to POWER THROUGH MIGHT

1. NAZI GERMANY
   The geographical development of Neitzschean and Fascist philosophy, adopted by Hitler with appalling consequences.
2. AMAZONAS
   Empty land must be claimed and occupied; it is vital for national development and the focus for a war.
3. SOUTH AFRICA AND NAMIBIA
   The geographical consequences of a state's domination by an elite group.
4. THE WESTERN FRONT IN THE FIRST WORLD WAR
   The consequences of individual and group aggression.

*The aggressive use of power, by individuals, states, and groups within states, has important effects on human geography. In return, geography restrains and facilitates that use of power.*

Chapter two

# Power through right

**Introduction**

The feeling of a right to some piece of land is a feeling which mobilizes and motivates people to act decisively. If they sense they are being deprived of ownership or control of what they see as their land, people can normally be persuaded to argue and fight for it – indeed they may need no persuasion. In this way, the relationship between people and territory (the man–land relationship) can strengthen political power by mobilizing popular support, especially where the sense of belonging is strongest, at the level of the state or province. It can deliver public opinion and recruit volunteer armies; it can secure territory, create a delusion of security in a state which is actually weak, and it can destroy a state where the right of possession is split. The aim of this chapter is to examine the man–land relationship which is the basis of this 'power through right'.

First, the chapter introduces the philosophy of power through right which considers the debate about whether or not it is natural for man to own and control territory, with reference to Rousseau and Locke. Unable to resolve the problem in this way, it moves on to consider the known facts about the way children and people in primitive societies feel about territory. This leads to the examination of an alternative, the Amerindian philosophy of the relationship between man and land.

The second part of the chapter moves away from the philosophy of the man–land relationship to consider its role in the partition of the world into territorial states. It shows how, for economic and cultural reasons, state territory (as opposed to any other sort of territory) became an important focus of activity, attachment, and culture.

There follow four case-studies, each of which illustrates one of the four salient points about the man–land relationship and its effects. The Holy Land (case-study 2.1) provides an example of a strong relationship between a group of people and one piece of territory; it analyses the historical and cultural relationship and some of the special problems it has caused. The case of the Falklands (case-study 2.2) provides an example of attachment to – and war over – a relatively worthless piece of territory: the Argentine claim is based on the expansionist geopolitical ideas of a state which considers its greatness thwarted, whereas the British claim, when threatened with the islands' loss in 1982, metamorphosed from a dubious legal one to a powerful ethnic one. Grenada (case-study 2.3) illustrates the danger of the illusions created by the feeling of a right to self-determination in an 'independent state', ignoring the facts of the inequality of states: the strength of attachment by Grenadians to Grenada may have been as great as the American attachment to the United States, but the latter is a thousand times more powerful. Finally, the case of Northern Ireland (case-study 2.4) is an example of how two historical and cultural claims to the same territory have evolved and discusses how they might be resolved.

Overall, this chapter illustrates the significance of a sense of right as a source of power. It mobilizes people and motivates them to fight battles they might otherwise have shunned. The man–land relationship plays a key part in shaping the world's political geography and it represents, therefore, a crucial aspect of the relationship between geography and power illustrated here by four case-studies of the world's trouble spots.

## The philosophy of power through right

### Ownership of territory

The most ancient of all societies felt no rights of ownership. The human family had no possession and no land. Attachments lasted as long as children were dependent on their parents and then all bonds were broken and individuals were free again. Rousseau's 'Noble Savage' lost his innocence only when the idea of owning property took hold. In his *Discourse on the Origin of Inequality*, Rousseau said 'the first man who, having enclosed a piece of ground, bethought himself of saying "This is mine" and found people simple enough to believe him, was the true founder of civil society' (Rousseau 1968b: 192). He went on to explain that the ownership of property, not just land but all property, was the root of most of the world's problems:

> from the moment one man began to stand in need of the help of another; from the moment it appeared advantageous to any one

51

man to have enough provisions for two, equality disappeared, property was introduced, work became indispensable, and vast forests became smiling fields which man had to water with the sweat of his brow, and where slavery and misery were soon seen to germinate and grow up with the crops.

(Rousseau 1968b: 199)

Rousseau considered that owning property was unnatural. He did not accept that ancient man had the same sort of territorial instincts as animals. On the contrary, the Noble Savage claimed nothing as his own and the first person who did claim possessions threatened everyone else by depriving them of some small part of what they needed collectively to support themselves. Inevitably, the response to this threat was that others also claimed some property as their own and, since this now had to be accepted in society, Rousseau devoted the *Social Contract* to the equal distribution of property (Rousseau 1968a).

Not everyone agreed that property ownership was unnatural. Locke's *Second Treatise of Government* gave the opposite view. God gave the world to men in common, but even the savage who never realizes he owns anything must have a claim which amounts to ownership on what he is about to eat before he actually consumes it; there must come a point when the particular fruit or particular beast or part of a beast is his and no one else's: 'He that is nourished by the acorns he picked up under an oak, or apples he gathered from the trees in the wood, has certainly appropriated them to himself' (Locke 1966: 15–16).

After this natural start, Locke introduced a theological argument. This was that God must have given man dominion over the world so that he could use it for his own benefit, in other words to the industrious who would cultivate it and use it profitably rather than to the idle or foolish:

God . . . commanded man also to labour, and the penury of his condition required it of him. God and his reason commanded him to subdue the earth, i.e. to improve it for the benefit of life, and therein lay out something upon it that was his own, his labour. He that, in obedience to this command of God, subdued, tilled, and sowed any part of it, thereby annexed to it something that was his property, which another had no title to, nor could without injury take from him.

(Locke 1966: 17–18)

In the days when Locke and Rousseau were writing, land was the one type of property that really mattered, and whether or not land ownership, like other sorts of property ownership, was a natural or cultural attribute of humanity became one of those long-standing and probably insoluble nature–nurture debates. In the case of land, the argument that the desire for ownership was natural always proved somewhat easier to show than for property as a whole because so much in nature was plainly territorial in behaviour. There followed the inevitable arid argument about whether human territoriality was based on the same instincts and origins as animal territoriality. In practice, animal territory is certainly not treated as property involving an abstract right of possession, exchange, and sale, let alone in the abstract without occupation, and so the analogy was never helpful. But there is also another very important reason why the analogy could never work; this arises from the unique two-way characteristic of the man–land relationship; it is not just that people feel that they own some piece of land, it is also that people feel they belong to the land. They feel a sense of attachment which goes beyond the one-way idea of ownership. It is this attachment which is the focus of the ideas and case-studies of the rest of this chapter. It is this sense of attachment which motivates otherwise tranquil people to fight to defend or to recover land either which they consider they own or to which they feel they belong; they will fight for land and covet it, they will yearn for land and die for it. To understand these kinds of feelings it is useful at this stage to digress from mainstream philosophy to look at models of child development by Piaget and others and at aspects of the philosophy of the North American Indian.

Attachment to territory

Important clues about why the attachment by people to land can be so strong are to be found in the literature on child development, and in particular in Piaget's *The Child's Conception of Space* (Piaget and Inhelder 1956). The child's mixture of subjective and objective as described by Piaget, the mixture of emotion and fact in the child's relationship with his environment, is never lost by adults. The adult adds more sophisticated concepts to the child's but he never loses the childish subjective and emotional view of his surroundings; on the contrary, he applies this view more widely, even to entire communities and states.

Many of the very earliest developments in a child's perception are important for understanding the nature of the man–land relationship. This is again because the emotional links between the individual and his familiar environment are additive. This means that they are not replaced through life by different emotions: the strong emotional overtone in environmental perception is always there, even when more complex emotions and more complex use of symbols are added. So the whole early development of perception shows how important the control of familiar territory becomes to everyone. Piaget highlights the child's need to feel safe in his surroundings, for which some sort of control, whether it be by individual, family, or community, is essential. This need is demonstrably carried into adult life.

During his first two years, the child gradually develops a clear and stable perception of his immediate surroundings and of himself as a physical entity separate from the rest of the world. At this stage, the objective and subjective worlds are quite mixed up as he projects his emotions on to his surroundings. Objects may be friendly; rooms may be scary. When this subjective way of seeing the world is carried through into adult life, it may develop into a belief in animism, myths, and magic; or it may remain a simple subjectivity in the adult perception of familiar and homely surroundings – the familiar becomes friendly, comfortable, and secure. When Rousseau's Noble Savage used his own labour to till his own plot of land, it was surely this subjective perception that told him the plot was not just the source of some of his food, but actually his own land, which he could shape and on which he depended; if necessary, he would have to defend it. The Noble Savage felt the same attachment to his environment as every child and every adult ever since.

Near to birth, objects do not exist to the child, he cannot see as adults do and has no adult sense of space. During the first two years, however, the child becomes aware of specifically territorial concepts; these are propinquity, proximity, and separation. He knows what is close and at the same time feels close. Still within the first two years, he also learns the ideas of enclosure and continuity. He is still highly egocentric; he thinks of things only in terms of his own relationship with them, so the context for his experience of enclosure and continuity is restricted to his own territorial safety and his own territorial threat, closeness, and separation – his own first strong feelings about his surroundings.

The child's relationship with his environment is closer to that of adults in societies which have not undergone a European or Asian type of development, and it is therefore helpful at this stage to look at the undeveloped or primitive man–land relationship. There are a whole variety of feelings of attachment in developed societies, ranging from communities who feel they belong to distant territories to individuals who feel violated at intruders into their half-acre of suburbia, but these are the consequences of several thousand years of social, economic, political, and cultural change of the sort that has taken place in Europe and to a lesser extent in Asia. To understand more of the root of the relationship between man and land, it is best to shift the focus from the child, not straight to this sophisticated modern world, but to the adult in more primitive societies. In Aboriginal and North American Indian societies, for example, land is not associated with individuals, but still the infant feelings of propinquity, proximity, separation, enclosure, and continuity all play their part in the man–land relationship.

In traditional Aboriginal society, the man–land relationship determines life and death; it is a strongly emotional one. Strehlow has made a major study of the relationship between Aborigines and their land. The Aborigine clings to his land with every fibre of his being; mountains, creeks, and water-holes are not just interesting or exotic landscape features but the work of revered ancestors who guard them. The whole countryside is his living, age-old family tree (Strehlow 1978). Eliade in her study of Australian religions (Eliade 1973) also identifies special landscape features as having emotional meaning, such as life-giving rivers that contain certain animals, guardian rocks, and sacred hills.

In the North American Indian tradition, land is viewed as a sacred object; it is a part of the system of life and it is not possible to own it. Mooney quoted the reply by Chief Smohalla when advised to use his land to dig for minerals and plant crops for sale for cash:

> Shall I dig under her skin for bones? Then when I die I cannot enter her body to be born again. You ask me to cut grass and make hay and sell it and be rich like white men. But how dare I cut off my mother's hair?
>
> (Mooney 1973: 716)

Again, in Hopi Indian families when a child was born, his Corn-Mother, a perfectly formed ear of corn, was placed beside him. The child was washed and rubbed down with corn oil. His aunts each gave the child a name and other relatives selected one of them; whichever aunt had originally selected the one chosen by the relatives then took the Corn-Mother and became the child's earthly godmother. The land was the child's spiritual parent.

Hopi Indians had parents and a clan, but their universal parents were the Sun-Father and Earth-Mother. Everyone was born from the Earth-Mother. The Corn-Mother tradition reflected this, since corn represented the earth and was spiritually far more important than the child's earthly parents. The Hopi creation myth maintained that the earth and human bodies were created in the same way; the axis running through human bodies, the backbone, corresponded to the axis of the earth, reflecting the inseparability of man from earth:

The Great Spirit is our Father, but the Earth is our Mother. She nourishes us; that which we put into the ground she returns to us, and healing plants she gives us likewise. If we are wounded, we go to our Mother and seek to lay our wounded part against her to be healed.

(McLuhan 1971: 22)

As the North American Indian and European concepts of land and ownership collided, the Indian view was written down formally for the first time and the incompatibility of the two approaches soon became obvious. A holy woman of the Wintu tribe of California expressed the feeling well:

We don't chop down trees. We only use dead wood. But the White people plow up the ground, pull down trees, kill everything. The tree says, 'Don't. I am sore. Don't hurt me'. But they chop it down and cut it up. The spirit of the land hates them . . . they blast rocks and scatter them on the ground. The rock says, 'Don't. You are hurting me'. But the White people pay no attention. How can the Spirit of the Earth like the White man? Everywhere the White man has touched it, it is sore.

(McLuhan 1971: 15)

In 1855, at a conference to organize the allocation of reservations for various Indian tribes, the Chief of the Cayuse expressed the Indian feeling on land when he opposed the European concept of landholding implied in the idea of the Indian reservation. The land's importance was religious. The land was given to people by the Great Spirit. They had no right to sell it.

The ground says, It is the Great Spirit that placed me here. The Great Spirit tells me to take care of the Indians, to feed them aright . . . . The water says the same thing . . . . The same way the ground says, It was from me man was made.

(McLuhan 1971: 8)

Thunder-Travelling-To-Loftier-Mountain-Heights, known as Chief Joseph, of the Nez-Perce tribe, expressed similar aversion to the European idea of possession and demarcation. To him the land simply could not be owned or ruled, in the European sense, by anyone.

The Earth was created with the assistance of the Sun, and it should be left as it was . . . . The country was made without lines of demarcation, and it is no man's business to divide it. The Earth and myself are at one mind. The measure of the land and the measure of our bodies are the same . . . [but] I never said

the land was mine to do with as I chose. The one who has the
right to dispose of it is the one who created it.

(McLuhan 1971: 54)

The cultural and economic development of European society made
North American Indian ideas incomprehensible, and conflict was
inevitable. It was not that the European desire to fence in land and
declare boundaries had any different origins from the Indian view,
just that European historical development had run another course.
The European and Asian ideas of the ownership of land, both private
ownership and communal belonging, were equally strong and just as
beautiful as that of the Earth-Mother. Europeans were capable not
just of dying to defend their own private property around which their
social and economic lives revolved, but they were also capable of a
grander vision of land quite as noble as that of the Aborigines and
the North American Indians although quite different from it. The
Jew would die for the Holy Land, the Irish Catholic and the Irish
Protestant would both struggle for Ulster and, in a more complex
political world, the Englishman would die for the Falklands and the
American for Grenada. From the Noble Savage until now, from the
new-born babe to the old soldier, man clings to the land, belongs to
the land, and is part of the land he calls his own.

## A focus for power through right: the territorial state

### The origins of the territorial state

The territorial state had no place in the primitive, ancient, and
medieval worlds. It was only in the later Middle Ages that people
in Asia and Europe started to feel that they belonged not just to their
neighbourhood and not just to a great world-wide community but to
something akin to the modern state. It was the result of a combination of man's emotional relationship with the land to which he feels
he belongs with the economic and cultural demands of the modern
world.

As society became more complex, some division of labour was
inevitable. There was soon a need for specialized farmers, merchants, shopkeepers, and so on. There evolved economic classes
each with a different vision of the social order and, inevitably, some
with antagonistic views. No longer was society seen by all its
members in the same light.

It soon became vital to reconcile different views or suppress
awkward beliefs because, without reconciliation or suppression,
arrangements for trade, commerce, and exchange could easily fall

into chaos. Some institution was needed with the power to overcome the antagonisms of civilized society. This power did not need to be great but it did need to be open-ended so it could use its authority to deal with unforeseen events and conflicts; such open-endedness was much simpler than to have a long list of separate items to come under its jurisdiction. Above all, it would have to have power which was visible and real, and it had to be endowed with the most basic attributes of objects, location, and extent. Land reified the state.

Territorial states took a long time to reach their modern form. Their gradual development can be traced from the rejection of primitive man–land relationship in Asia Minor and Southern Europe in favour of larger communities. It was part of that great change described by Homer and Aristotle, when several villages united into one community with some common interests to form the early city-states of ancient Greece and the river communities along the Nile and Indus and in Mesopotamia. Particularly in these latter cases, the need for territorial organization was made more pressing by a drying climate and the need to control limited resources.

It soon became obvious that straightforward armed control over territory beyond the city-state or river community itself yielded wealth with neither the cost of administration nor the complication of sharing it out. The tributes paid to the great empires of the ancient world were ample evidence of that; but it was not until the Middle Ages that the importance of a home-state with clear territorial extent acquired further economic significance.

In the Middle Ages, especially in Europe, economic power had come to be concentrated in four main forms: the control of the yield of the land (food and clothing); the control of passage (tollhouses on roads or at city gates); the monopoly privileged to produce or sell a particular commodity; and the right to mint money (thus manipulating at will the means of payment). Each of these four forms of wealth could not exist, amidst medieval insecurity, without the kind of political and military protection which could only be provided by a king or his barons. All the four forms of wealth also had a clear geographical basis. Each had to be located in specific areas and places, and rights had to be exercised within clear geographical boundaries. The boundaries could shift over time but still they were effectively state boundaries.

The rise of capitalism consolidated the territorial state. Capitalism needed a particular size of community. The city was too small; great multinational communities like the Holy Roman Empire were too unwieldy and uncontrollable. Capitalism needed large and mobile labour pools and free trade over wide areas with centrally organized infrastructure, especially for transport. The co-ordination of economic functions soon became one of the most important purposes of the territorial state.

## The cultural transition to the territorial state

Despite the economic importance of a degree of territorial control, the economic benefits of dividing the world into state territories are not a sufficient explanation for the dramatic changes entailed. There must therefore be more than an economic cause of this modern division of the world into states, and certainly more than an economic motivation behind people's attachment to the territory of the state where they live. For a fuller explanation, it is useful to turn to Benedict Anderson's incisive cultural analysis (Anderson 1983). He argues that the multinational religious community and the multinational dynastic realm kept the economic demand for a territorial state at bay until new cultural reasons for such a development caught up with the economic ones.

The three great religious communities of the medieval world were Christendom, Ummah Islam, and the Middle Kingdom of China. Most important of all, throughout the territories where one community had influence, one common language was shared. The script was of central importance: Latin, Arabic, or Examination Chinese. The Chinese approved of barbarians who learnt their script and the Koran was never translated from the Arabic until the last century. These were truth-languages. The culture associated with them overrode nationalism and localism. They enabled an Englishman to become Pope and a Manchu to become Emperor.

When he stresses the importance of language, Anderson admits that literates were only tiny reefs on vast oceans of illiteracy, but their importance lay in their centrality. Whereas, for example, lawyers' or doctors' languages today exist on the fringe of their society, the truth-languages and the culture associated with them were near the apex of their society. Society was seen as a strict cosmic hierarchy with the divinity set at the actual apex and the literates very near the top. Society was centripetal and hierarchical rather than boundary-orientated and horizontal with lots of equal functional groups.

A centripetal society with one truth-language was only sustainable while other languages had no other significant literature. It was unsustainable, for example, in Europe by the sixteenth century when the vernacular became much more important than Latin. Between 1600 and 1650 the Reformation ensured that Latin shifted from accounting for 80 per cent of texts to being a printed language for only a small minority. Language was territorialized. The truth-language was no more.

At the same time, people's knowledge of the world around them was widening fast. Auerbach (1953) described the disturbing impact which new cultural and geographical horizons had on Europeans in the sixteenth century. They were suddenly confronted with other possible forms of equivalent human life. They were faced with other religions. The existence of the Moslem world and the Chinese Confucian world implied splitting faiths up among territories; of course this could never be admitted by those at the top of the hierarchy, but it was increasingly undeniable. No faith could claim the world; each now had to define its territory and to secure it.

At the same time as the great religious communities were collapsing, the great dynastic realms also went into decay. In medieval times, the dynastic realm was the only political system people could imagine. It is very difficult to comprehend nowadays because serious monarchy is opposite to modern conceptions of life. Medieval monarchy was serious; it organized everything around itself at the centre and its legitimacy derived from God and not from the subject population, whereas in modern times, state sovereignty is fully and evenly operative over the whole of a demarcated territory and this sort of arrangement requires some legitimacy derived from the population or it is inoperable. In medieval times, political communities were defined by their centres and not by their porous and indistinct boundaries. The medieval dynastic realm did not even need to be contiguous let alone homogeneous. For example, the Holy Roman Empire was made up of lots of diverse places brought together by sexual politics – from Spain to Hungary, from Tuscany to Silesia. Nationality was unimportant in the medieval dynastic realm; there was rarely an Englishman ruling in London or a Frenchman ruling in Paris.

It was in the seventeenth century that the automatic legitimacy of a holy monarchy really evaporated. The beheading of Charles Stuart and his replacement by a plebeian Protector was a decisive shock. It is true that the Bourbons and even the later Stuarts still cured the sick by the laying-on of hands and behaved superficially like medieval monarchs; but principles of legality and constitutional rights were becoming the refuge of monarchy as their ancient rights withered away. Even in the east, the Son of Heaven became Emperor in the nineteenth century and the Prince of Siam came to study state constitutional monarchy in the west. In the west itself, evidence for the decisive shift to territorial statehood could be found in new armies like that of Prussia, which had been staffed by men of at least a dozen nationalities in the eighteenth century under Fredrick the Great but by Prussians alone by the middle of the nineteenth century.

The disappearance of the great religious communities and dynastic realms also brought to an end their concepts of time and place. A territorial state needs to have a clear locus in time and place, which people simply did not possess in the medieval world and before.

It is very difficult now to imagine any different concepts of historical time, but it was not a bit strange in medieval times to portray Christ and Mary in contemporary clothes. Medieval painters were not ignorant, they knew what they were doing, but what would now seem odd or blasphemous was normal to them. Again, as Auerbach (1953) explains in *Mimesis*, the modern Christian tendency to trace a long historical thread running from the Sacrifice of Isaac to the Sacrifice of Christ on to modern Holy Communion would be bizarre to a medieval Christian, to whom the gaps between prefiguring, fulfilment, and re-enactment were quite meaningless. To him it was all part of one picture visible at one time.

The concept of place was also quite different from the modern concept. Patriarchs or the Apostles would be painted in scenes taken from the countryside of Burgundy without it seeming either ridiculous or worthy of comment. Yet neither was this done through ignorance; it was known that ancient Judaea was different and looked different, but this was irrelevant knowledge. Belonging to a particular place in the world and belonging to a particular time in history were unimportant compared with belonging to religious and dynastic hierarchies powerful enough to create their own dimensions in the human mind.

The medieval way of thinking about time and place was hopelessly lost as the old hierarchies died. Exploration of the world and faster and more reliable communications, first through the printed word and later through improvements in transport, contributed to a new set of certainties and beliefs based on a new understanding of time dominated by the calendar and the clock and a new understanding of relative locational space. Man saw himself now as an organism moving through time in a particular place. In his efforts to assimilate this new set of ideas and in his search for new certainties, he discovered a new concept of community with shared destiny, shared salvations, not based on allegiance to a hierarchy, but based on similarity of time and place. The similarity also included shared language and shared security. The shared language was initially the more important in helping to create new communities clearly defined by history and geography. Hegel said that national newspapers replaced morning prayers. Quite so. Man had entered on a new era when he could say: 'This is my time' and 'This is my land'.

*Case-study 2.1: the Holy Land*

Christians, Jews, and Moslems: contrasting claims

The Holy Land is one of the best examples of conflicting claims of the right to possess territory. All through history, the Holy Land has been the home of several different cultures, races, and religions. These conflicts in the territory itself have also been of burning importance to many people outside the area altogether. Attachment by diaspora nations to their former territory is not unusual, but who controls the Holy Land is so important for so many people around the world that it can affect international relations and threaten world peace.

Of the three religious groups – Moslem, Christian and Jewish – with an interest in the Holy Land, the Moslem is the easiest to define. For Moslems, the issue is not the Holy Land of Palestine so much as the holy city of Jerusalem. According to Moslem belief the Temple of Solomon in Jerusalem was miraculously built around the rock of Abraham's sacrifice; and it was from that rock, which still bears his footprint, that the Prophet Mohammed ascended to heaven.

For Christians, the land as a whole is their Holy Land, scene of the earthly life of Jesus Christ. Christians have rarely wanted to live there, but they have always sought access for pilgrims and have supervised the sanctity of the many Christian holy places. Only the Crusades briefly sought control.

For Jews, the Holy Land is the Promised Land. The promise was contained in God's covenant with the Jews made on Mount Sinai and it has led to an intense feeling of belonging. Only for the Jews is the possession and occupation of land a central part of religion. Exiled by the Babylonians and later dispersed by the Romans, the idea of return to the Promised Land is central to the Jewish faith.

The aim of Islam is the submission of all people to the will of Allah; the Koran is not the history of a people but a guide to life. The aim of Christianity is Salvation in Christ, which is not restricted to any place or group; the New Testament is equally at home in Greece and Asia Minor as in the Holy Land. Judaism is different; its focus is the divine revelation of a way of life to be lived by the Jews in the Holy Land.

The Jews: a historical and theological right

Islam and Christianity put a lot of stress on a relationship with God in preparation for future life. Judaism has very little concern with the hereafter; Jewish law is primarily concerned with community life in this world. Jewish law is concerned with the intimate details of the domestic, social, and commercial life of the Jewish community in the Holy Land as well as their religion and relationship with God. The Old Testament is the history of the Holy Land and Jewish festivals reflect its political and social history; for example, the agricultural festivals follow the Holy Land's agricultural seasons, and at the Feast of Passover, Jews pray to return to the Holy Land: 'Now we are here, but next year may we be in the land of Israel. Now we are slaves, but next year may we be free men. Next year in Jerusalem!'

Nowadays not many Jews are strict in their religion, but they still share an overwhelming feeling of right to possess the Holy Land and to live there if they wish. This feeling has been strengthened by the Jewish experience of second-class citizenship and frequent persecutions under Christian and Moslem rule. It has reinforced their sense of identity, not as a mere religion but as a people, a complete cultural entity just as much as, say, Spaniards or Turks, but a people in exile.

Jews always regarded their exile from the Holy Land as temporary. A return of their rights would simply restore the historical status quo ante. Throughout the whole history of their 2,000 years of exile, religious Jews struggled to the Holy Land to pray or to die and be buried on the Mount of Olives in Jerusalem; there has never been a time when the Holy Land has been without Jews. All other possible havens at times of persecution, such as Uganda and Argentina, were always refused. Only the Holy Land would do. Theologically, the Jewish exile from the Holy Land was a manifestation of the will and wrath of God and return would signify redemption. The return would presage the longed-for and ultimately inevitable coming of the age of the Messiah.

## The Jews: a political and cultural right

These historical and theological arguments were a heady mix from which there developed the profound political and cultural relationship between the Jews and the Holy Land. The Nazi Holocaust, the death of six million Jews, and subsequent gentile guilt, made the modern case for return irresistible. It was clearly put by David Ben-Gurion who was to become the first Prime Minister of independent Israel.

Ben-Gurion's case started from the premise that the Jews, 'like every other free and independent people' were entitled to fair treatment, including their own land. Jews were entitled to escape from dependence and discrimination.

Whenever you have two people, one a strong group, powerful, and the other weak and helpless, there is bound to be mischief. The strong group will always take advantage of the weaker group

. . . sometimes [the stronger group] are excellent people, but not always, and there is discrimination – not necessarily legal, political, or economic – sometimes merely a moral discrimination, and they [the Jews] do not like it, and they do not see how they can change the whole world, so they have decided to return to their own country, and be masters of their own destiny.

(Ben-Gurion 1947: 56–7)

Ben-Gurion also put the historical argument but in a modern political context: 'This always has been, and will remain our country. We are here as of right. We are not here on the strength of the Balfour Declaration or the Palestine Mandate. We were here long, long before' (ibid).

The exile was as yesterday. He used the analogy of the European Jews who emerged from concentration camps to find other people occupying their houses. In the case of the house of the Holy Land, they were not even occupying them completely. His analogy was with a house of fifty rooms the Jews had left and now they had come back to find five occupied by someone else and the rest neglected, and they wanted to come back only to the forty-five and leave the five (ibid).

Ben-Gurion also returned to the emotional attachment of the Jews to the Holy Land, the Jewish 'love for Zion'. He argued that this was no mere mystical conjecture, but real feeling expressed by real migration. There was no desire to conquer the Holy Land for power, but simply to be there because it was in the heart of every Jew. Jews had failed to settle on fertile lands in Russia, America, and Argentina, but they had settled in Israel, despite great difficulties, because of their love for the land.

We here [in Palestine] are the freest Jews in the world. Not in a legal sense. On the contrary, here we are deprived even of equality before the law. We are living under a most arbitrary regime . . . . In spite of all that, we here are the freest Jews in the world. Freedom begins at home; it begins in the human mind and the human spirit, and we are free men, and here we are building our Jewish freedom, more so than all other Jews in the entire world. Why? Why do we feel freer than any other Jews? Because we are self-made Jews, made by our country.

(Ben-Gurion 1947: 65–6)

The West Bank: a conflict of rights

Israel became the Jewish state in the Holy Land in 1948 with sovereignty over those parts where Jews were in the majority, as defined by the United Nations. The existence of the state was not accepted by the surrounding Arab states and the resulting war left Israel in control of about two-thirds of Palestine west of the river Jordan. The one-third which remained in Arab hands became known as the West Bank. In June 1967, an Arab invasion was defeated in the Six Day War and Israel occupied the Golan Heights, Sinai, and the West Bank. In November 1967, the United Nations Security Council adopted Resolution 242, 'emphasizing the inadmissibility of the acquisition of territory by war'. It called for Israeli withdrawal and Arab recognition of Israel. Neither took place, and in the 1973 Yom Kippur War the Arabs attacked again and were defeated again; the West Bank and other territories remained in Israeli hands.

The successful peace initiative which resulted in the Camp David Agreement, signed by Israel and Egypt in March 1979, failed to solve the problem of the West Bank. The eastern limit of Israeli control is still on the River Jordan. The West Bank has now become the primary focus of the struggle between Arab and Jew in the Holy Land. At the same time it has created a dilemma for the Jews; the West Bank is part of the Holy Land which they feel is theirs by right, the ancient biblical lands of Judaea and Samaria, but the West Bank population is almost entirely Arab despite controversial Jewish settlement. Ben-Gurion's 'empty house' argument does not apply to the West Bank.

The Camp David Agreement created alarm among Israeli national religious groups and settlers. They thought that the loss of Sinai could also happen to the West Bank. To pre-empt this, they encouraged more settlement building and more Jewish migration to the West Bank. They took the view that the number of Jews in Sinai was too small and drew the lesson for the West Bank. Some 10,000 Jews would not assume Israel's presence but 100,000 would. This kind of attitude led to the creation of countervailing 'peace' groups on the political left, reflecting a split in Israeli Jewish thinking.

Laffin (1979) and Oz (1983) have explored by interview and discussion the territorial and moral conflicts in Israeli Jewish society over the West Bank. Some questions are raised in Jewish minds about the long-run risk to the very existence of Israel involved in possessing land inhabited mainly by Arabs. Other Israeli Jews respect Arab democratic rights and have no wish to dominate any part of the Arab world or to treat anyone (including Arabs) as second-class citizens. Laffin and Oz illustrate a conflict between the Jewish feeling of belonging to the Holy Land, the Jewish moral imperative, and the perceived need for territorial control.

The Jewish feeling of belonging to the Holy Land and having a right to the whole land is as strong as ever among many Israeli

Jews. Laffin and Oz show that the full claim on the West Bank as rightfully Israeli Jewish territory is not just associated with the political right wing, but is a view shared by many working-class Jews of Sephardi (usually Middle Eastern) origin, some of whom have themselves been – or their parents have been – expelled from Arab lands. They see no particular strength in the birthright argument for Arabs and point out the size, number, and wealth of Arab states – the Arabs should leave or be made to leave. A good many young Jews also see West Bank Arabs as troublemakers with whom negotiations are really impossible, a view which has gained popularity since the 1988 West Bank riots. In their view, the Jewish right to the Holy Land means just that and there is no room for compromise.

The political left wing makes no claim on the West Bank and fears for the democracy of the state with over a million second-class citizens. These modern moral imperatives modify the feeling of territorial right among Jews from a more politically sophisticated European background. Many on the left take the simple view that the Arabs, like the Jews, should have control over that part of the Holy Land where they were born and where they form the majority.

In practice, both the strongly territorial view and the liberal moral approach are both superseded by the practical requirements of territorial control. Israeli control over the West Bank is better than any of the likely alternatives from the point of view of Israeli military security. Any Arab state in the West Bank is likely to be hostile. Some of the leaders of the Palestinian Arabs are still dubious about the Jewish state's right to exist at all. There is a modern state to defend and a people to protect. In this scenario, the historical and moral rights to territory on the West Bank and the practical consequences of widespread Jewish settlement are important bargaining counters in a harsh world in which theoretical territorial rights play second fiddle to the hard facts of territorial control.

It is perhaps remarkable that there are any hard facts of territorial control at all. The Jews were a weak people, persecuted and widely scattered. Some were wealthy and influential but the vast majority were poor and many were murdered by the Nazis. They derived their strength and their power from a certainty that the Holy Land was theirs and that one day they would return to it. It is a testimony to the strong reality of power through right that the landless and nearly destroyed race of 1945 is conducting an internal debate about the West Bank today.

*Case-study 2.2: the Falklands*

Introduction

Both Britain and Argentina have had serious doubts at various times in the past about their respective claims to the Falkland Islands. Certainly neither has an indisputable legal claim. On the contrary, such legal argument as there may be is used to back up the emotional claims of each side, incomprehensible in both cases to the other side. Argentina and Britain both feel they have a right to the Falklands based on natural justice. In the case of Argentina, it is a right derived from a sense of thwarted destiny on a continent pre-occupied with geopolitical ideas and where European and American involvement are always seen as threatening the continent's independence. In the case of Britain, it is a right derived from the ethnic links between the British and the Falkland Islanders, and the Falkland Islanders' local right of self-determination.

The legal arguments

The legal history which provides technical back-up for both sides in the Falklands dispute can be briefly stated. At first the islands were probably sighted by the English Captain Davis in 1592. In 1600 they were certainly seen by the Dutch, and later visited by the Spanish and French. They were claimed by the French in 1764 but were transferred to Spain in 1767 for £24,000; however, this arrangement was apparently unknown to the British who had meanwhile claimed the islands for themselves in 1765. Spanish protests were made in London and this seems to have been the start of the dispute. In 1773 the islands were abandoned by everybody with Britain and Spain still in dispute. In 1829 Buenos Aires claimed the islands, insisting that they had inherited the Spanish claim. Spain, however, had not yet even recognized Argentine independence and in fact only did so in 1857 when nothing was mentioned about the Falklands. Meanwhile, an Argentine settlement was established on West Falkland in 1829 and a British settlement on East Falkland in 1833.

Various attempts have been made to unravel the legal rights and wrongs. But to understand the way in which popular national claims have developed, it is far more useful to look at the development of geopolitical ideas in Argentina, British incomprehension of those ideas, and the alternative British geopolitical response.

The Argentine right to the Malvinas

For Argentina, the claim on the Falklands (Malvinas) is deeply rooted in geopolitical thinking. A good deal of literature was (and still is) devoted to proving Argentina's claim and relating the claim

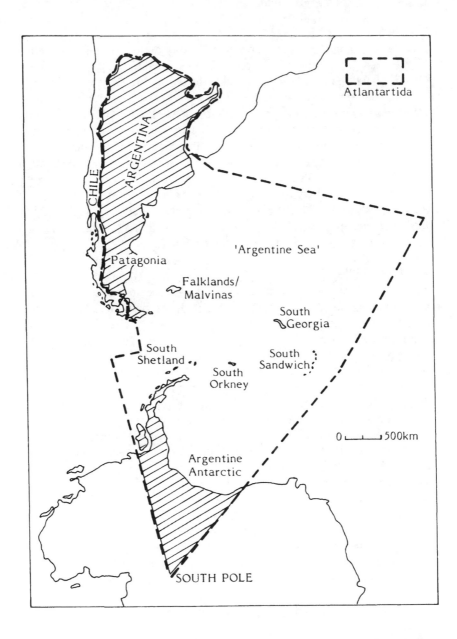

*Figure 2.1* Argentine concept of *la Atlantártida*

to Argentina's wider rights in the South Atlantic and the Antarctic; some of the literature specifically calls for military action (Child 1985: 117–22). The legal argument is rarely dwelt upon; the economic argument is not strong – even the most optimistic economic assessment of possible potential resources of gas, oil, and krill amount to nowhere near the costs of military campaigns or garrisons. The significance of the Falklands in Argentine geopolitical thought far outweighs any assessment of the islands' legal status or economic significance. It is not even the traditional geopolitical approach that saw the islands as a vital naval choke-point in the two world wars that is important in modern Argentine geopolitical thinking; it is a different kind of geopolitics, the analysis of what a state might be, the conscious enhancement and then exploitation of Argentine self-esteem that has persuaded ordinary Argentine people that Argentina has a right to the Falklands.

Foreigners tend always to underestimate Argentine strength of feeling about the Falklands; it is not restricted to any class or political group. Argentina is seen as always thwarted and always prevented from achieving greatness. There was obviously never a possibility of an empire in the European style, and opportunities for expansion inland soon ran into the sand of disputes with Chile over Patagonia and the Andean frontier. The one obvious way for Argentina to expand was into the South Atlantic and Antarctic, and furthermore this was an area of relatively soft targets, next to South America but no one's core sovereign territory. It was also an area of unknown resources and could be portrayed as a new frontier for exploration and greatness. Here was the opportunity to enhance the self-esteem of the Argentines and here was the opportunity for the minds of the Argentine people to think of future glory and not of present ills. Yet here were the British, already heavily involved in the Argentine economy, backed by the even more powerful Americans, trying to halt this harmless project, this best chance for Argentina to achieve greatness in one of the world's last great frontiers.

The virility of Argentina was to be put to the test in the South Atlantic and in the Antarctic. If the southern frontier were to be neglected, others would take the place of Argentina – the same Latin American geopolitical Law of Valuable Areas would apply as Brazil, Peru, and Ecuador used in the Amazonas (Chapter One, case-study 1.2). Indeed the empty quarters of Antarctica were already being allocated and explored with only a very small Argentine interest and the way to the South Atlantic and Antarctic Oceans was blocked by the British in the Falklands and other southern islands. Negotiations with the British were effectively blocked by the British insistence on the detailed citizenship rights (if not at this stage an actual power of

veto) for a mere 1,800 settlers, but at the same time there was every
diplomatic indication of a growing British desire to be rid of its
South Atlantic problem (Hansard Report 1980). The military regime
was in some economic and consequent political difficulties, and the
time was right for it to make use of the political ground it had been
preparing. President Galtieri's visit to Argentine bases in Antarctica
and an increasingly militant approach to the British over the
Falklands started to indicate a possible culmination in war.

It fell to Admiral Fernando Milia to make an unequivocal intellectual link between geopolitical ideas and military action (Milia 1978).
He described the vision of an expanded Argentina including *la Atlantártida*, the whole of the South Atlantic, the Falklands, and up to a
quarter of Antarctica, (Figure 2.1), and went on to explain the need
for a military solution. He developed the idea of *la Atlantártida* as
a coherent and integrated geopolitical space, and set out to show that
Argentina's future as a nation was unequivocally linked to the
knowledge, occupation, and use made of *la Atlantártida*. It was the
challenge of a new frontier that would test the strength and will of
Argentina and help her to develop that internal cohesion that
characterized a nation-state as opposed to a plain political state.
Milia never doubted that Argentina may well have to fight to fulfil
its South Atlantic and Antarctic destiny and should be prepared to
do so.

That was in 1978. The war again Britain was lost in 1982. In 1984
a leading Argentine newspaper, *La Nación*, carried the following
editorial:

> Regardless of the setback we have experienced, we must look
> into the future by viewing the events which have occurred as the
> contemporary stage of the century-old battle that Argentina has
> been waging against Great Britain.
>
> (Child 1985: 122)

It went on to say that Britain was motivated by the desire for
renewed imperial power and economic gain; the Argentine army was
getting ready again: *la Atlantártida* was not forgotten.

The idea that Britain and the United States were expanding their
imperial power politically and economically into the South Atlantic
has a great deal of credence in Argentina. The idea that a base is
being built related to NATO in the Falklands under British control
is widely accepted. It seems a plausible explanation to Argentines
why the British fought, why the United States backed the British,
and why a new military airport has been built. To the British, the
airport was simply a logistical necessity for maintaining its Falklands
garrison, but the Argentine view can easily be understood in the

context of Argentine geopolitics. To Argentina, the airport will inevitably become a key base from which the British and Americans will be able to project their power into *la Atlantártida*; imperial powers suppressing Argentina, and, as they have always sought to do, preventing her greatness.

## The British right to the Falklands

To the British, the Argentine view was quite incomprehensible. Until 1982, the Falklands were virtually unheard of in Britain. In so far as they were thought about by anyone at all, they were vaguely expected to take their turn near the end of the decolonization process. There were already close contacts between the population of the Falklands and Argentina; many were related to the large British population in Argentina (some thirty times larger than that of the Falklands) and there were many other personal and commercial links. London and Buenos Aires had been negotiating for some time over the future of the islands; in 1977 the Foreign Office minister responsible, Ted Rowlands, had recommended that a way should be found to allow the islanders to stay British but to give Argentina sovereignty over the territory. In 1980 the new minister, Nicholas Ridley, discussed fishing agreements and emigration from the islands and had some minor clashes with his Argentine counterpart over their sovereignty, for the British wanted the talks to keep to economic and industrial developments at this stage (Ridley 1980). Social matters such as education and health arrangements were to be discussed next, and eventually more contentious but less practical matters such as sovereignty. At that point, the Falklands were a purely technical issue for the British. For the Argentines, of course, they were anything but technical.

Argentina's insistence on raising the sovereignty issue was met with complaints that harping on this abstract matter simply reduced the economic well-being of the islands, stagnated the economy, and encouraged migration. In a parliamentary debate in December 1980 (Hansard Report 1980), the sovereignty issue brought the first geopolitical response from Britain. It was a response based on ethnicity. Sir Bernard Braine, for example, talked about the islanders as 'wholly British in blood and sentiment' (Hansard Report 1980: 129) and it was agreed that 1,800 people of British decent deserved British support. But at this time the main concern was with the economic development in the islands and the administrative problems and diplomatic details of reaching an agreement with Argentina.

Parliamentary debates in 1982, the year of the war, displayed a dramatic shift of emphasis from technical to ethnic. The publicity given to this political stress on the ethnic links between Britain and the Falklands quickly heightened public awareness. As the Argentine threat to the Falklands mounted, so it became clear that the British had a response to the Argentine geopolitics. The response was to be ethnicity, the application of British ethnic feeling to the right to own a place inhabited by people perceived as ethnically similar (Calvert 1983). In fact, the Falkland Islanders were economically different, socially different, politically different, and geographically utterly unrelated, yet they were perceived as ethnically similar. Such similarity as there was amounted in reality to little more than language and a rather vague idea about race. The Falkland Islands were seen nevertheless as territory which the British had the right to own virtually exclusively because people thought of as British lived there. At first, the government was surprised by the explosion of emotional support for the Falklands; later, government propaganda with media support set out to, and was able to, sustain a high level of emotion to create and maintain support for military action. *The Times* declared: 'The Falklanders are British . . . the Falkland Islands are British territory; when British territory is invaded, it is not just an invasion of our land but of our whole spirit. We are all Falklanders now' (*The Times* 1982). Margaret Thatcher talked on television of the impossibility of betraying 'our own people' and in parliament of the islanders who are 'British in stock and tradition' (Calvert 1983: 137).

The grandiose ideas of *la Atlantártida* used by the Argentine leadership and media to stir the Argentine people into a firm belief in the need to fight for their territorial right clearly had its counterpart in the British use of ethnicity. Broken treaties and affronted national pride would not have been enough to engender British popular support for war and loss of life in the South Atlantic. Unlike Argentina, Britain had few geopolitical ideas about the South Atlantic – at the most a vague notion of the value of a South Atlantic base and of Antarctic potential. The Argentine invasion forged a new geopolitical link between Britain and the Falklands dominated by the idea of ethnic solidarity. After the war, this simple link was generally put aside in favour of principles of peaceful self-determination and the non-use of force; but it was ethnicity that had provided that mobilizing motivating feeling of right at a time of crisis. It was ethnicity that bridged the oceans.

*Case-study 2.3: Grenada*

The illusion of self-determination

The idea of self-determination is mixed up with the idea of the state. Although the term can be applied to groups of people in particular regions of states or to people who are members of special groups within states, or even to a collectivity of states, the desire for self-determination, as a motivation for organized action, now only applies to states or to places thought of by a fair number of their inhabitants as potential states. Self-determination means people having the means to make independent political, social, and economic decisions, constrained only by voluntary international agreements. This is not something easily or commonly achieved. People can only really determine their own affairs and effectively own and control the land where they live if two preconditions are met. First, the state must be potentially economically viable in so far as it must not depend on other states for the basic needs of life. Second, it must be politically viable; its conduct must be such that it does not attract intervention under international law and such that it does not appear to threaten more powerful states within whose sphere of influence it lies. Therefore, self-determination does not necessarily apply realistically to all the states that claim it, particularly very small states.

The tendency to give equal status in international law to increasing numbers of territorial units, no matter how patently unequal in practice, has complicated international relations and opened up vulnerable people to exploitation and pressure, which, if they were not nominally in independent states, they could have avoided. The case of Grenada is discussed here. The rulers and many of the people of Grenada had the illusion of self-determination but were neither economically nor politically in a position to enjoy it. Their formal independence encouraged them to seek an unattainable level of real independence from the United States.

It is sometimes overlooked that there are alternatives to state independence which may offer greater real self-determination. Indeed, attachment to Britain allowed a few hundred Falkland Islanders unparalleled rights of self-determination and the same would probably be true if anything were to threaten Grenada's French Overseas *Département* neighbours, Guadeloupe and Martinique. It is likely that having large numbers of small states, all with their own theoretical rights and prerogatives, can only be a short-lived phase in political development. It is no way to run a highly unequal and highly aggressive world. The case of Grenada

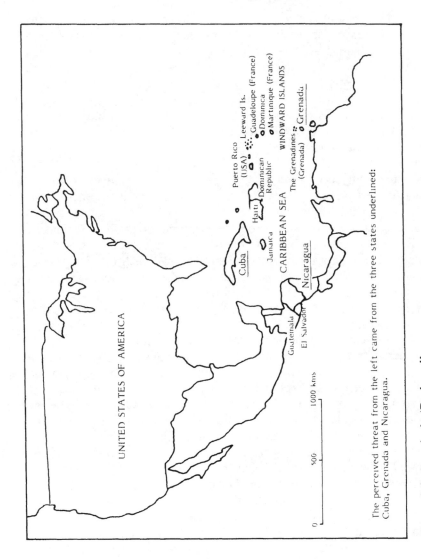

The perceived threat from the left came from the three states underlined:
Cuba, Grenada and Nicaragua.

*Figure 2.2* America's 'Backyard'

shows where the struggle for self-determination can lead in the face of more powerful superpower ideology and interests.

Grenada and the United States' sphere of influence

Grenada is a small island at the southern tip of the Windward Islands in the eastern Caribbean (Figure 2.2). Most of the 110,000 population are in agricultural villages dependent on cash crops such as cocoa, bananas, and nutmeg. Before the United States' intervention in October 1983, tourism was increasing at a level likely to supplant agriculture as the dominant economic sector.

Grenada became independent from Britain in 1974 and immediately came under the corrupt one-party rule of an eccentric leader, Eric Gairy. An opposition movement, the New Jewel Movement, was violently kept down until 1979 when it staged an almost bloodless and widely popular coup while Gairy was away at the United Nations (talking about flying saucers). The new government, which included socialists and Communists, was unacceptable to the United States.

The United States' economic and political power over the Caribbean was well known even before its rise to world superpower status, when it had applied the Monroe Doctrine of 1823. The Monroe Doctrine sought to remove all European influence in the region and supplant it with benign US protection. This was illustrated when, in 1898, US forces expelled Spain from Cuba and Puerto Rico. *The New York Times* commented: 'There can be no question of the wisdom of taking and holding Puerto Rico . . . . We need it as a station in the Great American Archipelago, misnamed the West Indies' (Latin American Bureau 1982: 10). The Caribbean was to be a part of the United States' greater destiny. Its increase in military potential was related to the dramatic growth of American economic interests in the region, especially the completion of the Panama Canal in 1914.

By the end of the Second World War, the emergence of a bloc ideologically opposed to the United States promoted a new agenda for subsequent American involvement overseas. The foundation of post-war US foreign policy was opposition to Communism; every administration since 1945 has permitted direct or covert intervention to forestall Soviet Communist ambitions, real or imagined. States near or bordering on the United States have been seen as buffer states in which the United States is committed to intervening directly if necessary. Since 1945, for example, US forces have directly intervened in Guatemala (1954), Cuba (1961), Panama (1964), and the Dominican Republic (1965), with El Salvador and Honduras receiving substantial military aid and advice. In 1965,

President Johnson justified the invasion of the Dominican Republic: 'The American nations cannot, must not, and will not permit the establishment of another Communist government in the Western hemisphere . . . . Our goal is in keeping with the great principles of the inter-American system' (Latin American Bureau 1982: 64).

The Grenada intervention was also to be portrayed in this way. After the Vietnam War, Presidents Nixon and Carter removed direct action from US foreign policy. Communism was to be opposed only with military aid and not with soldiers. There was also an overt use of economic aid and influence as a means of intervention. Michael Manley, for example, accused the Carter administration of manipulating debt relief to produce an austerity package that would ruin his 'socialist' programme in Jamaica.

### The Grenadian revolution and the United States

In March 1979, the United States was confronted with the Grenadian revolution. A message was delivered at once to the new Prime Minister, Maurice Bishop, by the US Ambassador, warning the new government about the possible consequences of conducting a foreign policy of which the United States might disapprove, for example the development of closer ties with Cuba. Bishop's reply was blunt and contained the illusion of self-determination in foreign affairs and other matters typical of small independent states in both its content and its consequences. It concluded: 'We are not in anybody's backyard and we are definitely not for sale. Anybody who thinks they can bully us clearly has no understanding, idea, or clue as to what material we are made of' (Latin American Bureau 1984: 46).

The Reagan administration came to power in the United States in January 1981; post-Vietnam paralysis was over and direct intervention was back on the agenda of US foreign policy. The new administration saw world affairs as a fundamental ideological confrontation. The softening of US foreign policy after the Vietnam War had encouraged Soviet-Cuban adventurism, seeking to undermine US interests and isolate the west. The Reagan administration would aim to respond by flaunting its military capability to restore respect from the Soviet Union and its allies. In addition, there was one quite distinctive element in the Reagan doctrine, the start of an anti-Communist offensive to destroy any regime ideologically opposed the the United States, particularly in the Americas. Defense Secretary Weinberger announced the offensive on El Salvador, Nicaragua, and Grenada in April 1981, insisting that the United States should be prepared to halt and seek to reverse the geographic expansion of Soviet control and presence, particularly when it threatened a vital interest or further eroded the geo-strategic position of the United States and its allies (Klare 1983).

In the Reagan doctrine's world of simple ideological divisions, Grenada's defiant left-wing regime was a threat. It was also a military threat; Grenada's location at the southern tip of the Caribbean basin could form a complementary gatepost to Cuba in the north, to enable the Soviet Union to gain a denial capability throughout the entire region if they could militarize both Cuba and Grenada (Figure 2.2). This military potential would in turn enhance Grenada's self-confidence in attempting to export its revolution to neighbouring small states.

### The destruction of the Grenadian revolution

Grenada presented an opportunity for the United States to put its new offensive into effect. A campaign of destabilization against Grenada was started. First, the State Department in Washington eroded Grenada's tourism earnings by warning travel agents that Grenada was not a safe destination. Second, the Reagan administration denied Grenada the finances for the construction of a new airport. Then, throughout 1981, the United States increased its naval presence around Grenada, ostensibly to carry out manoeuvres. As part of these exercises, US forces staged a mock invasion of an island group fictionally named 'Amber and the Amberines', a deliberate play on Grenada's colonial title 'Grenada and the Grenadines'. By March 1983, the US Navy had established an apparently permanent presence six miles off the coast of Grenada, which the Bishop regime construed as intimidation and a deliberate attempt to provoke an irrational response.

In October 1983, there was a coup in Grenada which provided the excuse for direct intervention under international law. The Marxist group in the New Jewel Movement, led by Bernard Coard, took over from Bishop and executed him. This was followed by some rioting and an appeal by the Governor-General, a constitutional figure-head, for outside help. On 23 October, Eugenia Charles, Prime Minister of neighbouring Dominica, also made a specific appeal to the United States to invade Grenada under the terms of the Treaty of the Organization of East Caribbean States. In addition, the United States expressed concern for some of its citizens in Grenada, medical students who turned out to be anxious but un-threatened and unharmed (Gilmore 1984).

Although international law has a statist basis, it also provides for intervention on certain occasions. It aims to protect state sovereignty, legal equality between states, territorial integrity, and the integrity of domestic jurisdiction, but intervention is allowed if it is to preserve international peace and security, to protect human rights, or to ensure that self-determination itself is preserved (Cassese 1986). Although the primary responsibility for ensuring peace and security lies with the United Nations Security Council, individual members can also act where danger threatens, as the United States clearly thought was the case in the eastern Caribbean. Second, the United States claimed that it acted to protect its own nationals, the medical students, and the Grenadian people as a whole from the abuses of human rights being perpetrated by the new government; in this it had the support of the Governor-General, who actually requested military intervention. This was linked to the third reason allowed for military intervention in another state under international law: the loss of self-determination by Grenadians following the coup. Certainly, Grenada had an un-democratic and unelected regime under Bernard Coard, although there is no reason to assume that the Governor-General was any more representative of Grenadian opinion – indeed he was one of the few politicians left over from Gairy's days and was also unelected.

The Reagan doctrine was enforced on 25 October 1983, when 2,000 marines and airborne rangers seized Grenada's airport and capital. The presence of armed Cuban construction workers led to an escalation in hostilities. By 28 October there were 6,000 US troops on Grenada, together with some 500 members of other Caribbean armed forces.

### Conclusion

The delusion of independence created by a statist world system is bound to lead to situations like that of Grenada in 1983, when legally independent states claim a right to a level of control over their land and their destiny that inevitably brings them into conflict with more powerful neighbours and with the superpowers. The rigidity of statism allows disputes to go so far as to make military intervention the only way a state motivated by the reality of power through might can bring about the downfall of a regime streng-thened by its conviction of power through right.

### *Case-study 2.4: Northern Ireland*

### The two claims to Northern Ireland

Neither opposing faction in Northern Ireland identifies with the territory of Northern Ireland. Each has a competing claim to live in, to own, and to control the territory of Northern Ireland, but it is also the aim of each party that Northern Ireland should be part of a neighbouring state, either the Republic of Ireland or the United Kingdom. Both look to their interpretation of the history of Northern Ireland to support their claim. The gist of the Roman Catholic Irish argument is a right based on original occupation and a concept of the physical unity of the island of Ireland. The gist of the Protestant Ulstermen's argument is that they form the majority of the population in Northern Ireland and their wish to be part of the United Kingdom should therefore be respected, which would amount to self-determination by the majority in this province which has existed as a political entity for longer than many a state where self-determination is regarded as an indisputable right.

The historical and cultural background to the two claims

The first organized involvement from across the Irish Sea in the affairs of Ireland was an Anglo-Norman invasion in the twelfth century, but this and some minor subsequent incursions only ever resulted in England's partial and temporary political control. Far from being assimilated into anything like English culture, the Irish assimilated the English colonizers and their offspring into Irish culture. Through the five centuries which followed, however, Ireland was successfully subdued by the English militarily and politically. The northern part of Ireland, Ulster, was the last and most difficult to subdue. So during the seventeenth century, the English authorities encouraged a 'Plantation' of settlers from England and Scotland. They would be a loyal and politically stable population. But the 'Plantation' neither displaced nor assimilated the indigenous Irish who became dispossessed and marginal tenants, beggars, or manual labourers, while the settlers enjoyed a privileged economic and social position (Buchanan 1982).

In disputes between the two populations of Ulster that occurred up to roughly the end of the eighteenth century, mainly concerning title to land, the language difference was the focus for taking sides, the symbol of difference. From the nineteenth century onwards, religion was the symbol; by this stage most disputes centred on trade, urban, and industrial matters, and it was religion that was used for seeking government support from London or solidarity with neighbours to the south. In other places, Protestants and Roman Catholics coexisted peacefully; although from a doctrinal point of view each might have regarded the other as heathen, as simple-minded followers of the Pope, or as unfortunates who had lost the true faith. But in Northern Ireland, religion is symbolic of profound cultural differences.

It is unusual for the symbol of cultural difference to shift from language to religion as it did in Northern Ireland. The possession of land seems to be the key, since it happens where a cultural or ethnic group is deprived of territory. Religion endures lack of territory much more easily than language; it is nearly always possible to carry on religious practices separate from most aspects of daily life, but language is needed to get on with others in society. The Jews retained their religion but only the religious part of their Hebrew language until they had their own state again; the same thing is happening now to Tibetans in exile, where generally the second generation of exiles speak only poor Tibetan but are diligent students of Tibetan Buddhism. Ulster's Roman Catholics lost their land and had to use the English language in everyday life, while a knowledge of Gaelic became something of a luxury. Without control over territory, their Roman Catholic religion became the symbol of their Irish culture (McAllister 1983).

It is easy to miss the symbolic nature of the religious divide and consequently to see the relationship between culture and territory in Ireland in mainly religious terms. The British in the twentieth century have tended to look on both Protestant and Roman Catholics as essentially Irish with religious differences, and have missed the point about the symbolism. They have consistently failed to recognize the depth of the cultural divide that lies behind the symbolism: two interpretations of history, two cultures which mix only at the most superficial level, two senses of superiority and of oppression, and two competing and exclusive territorial claims. One of the best-known and most glaring examples of this mistake was a speech made in Belfast in 1921 by King George V, the very year when Ulster was separated from the rest of Ireland, in which he appealed to Irishmen 'to stretch out a hand of forbearance and reconciliation, to forgive and forget, and to join in making for the land they love a new era of peace, contentment and goodwill!' (Nicholson 1952: 460).

Barritt and Carter (1962) made an important study of Belfast in 1972, in which they gathered a great deal of information about Protestant and Roman Catholic attitudes and patterns of life. They helped to identify more precisely than had been done before what lay behind the symbolism of the religious divide. They confirmed an almost total absence of social exchange (including intermarriage) and economic exchange only rarely on equal terms. The two communities led separate cultural lives emphasized by their children's attendance at separate primary and secondary schools with separate syllabuses. This educational divide was particularly important, confirming for children in the wider domain the attitudes of the home. Although the teaching language was the same in Protestant and Roman Catholic schools, the basic orientation was different. In Protestant schools, Irish history was taught as an incidental part of British history, but in Roman Catholic schools, Irish history and the Gaelic language were both recognized subjects in the junior and senior curricula. Under pressure from the EEC and British governments since Barritt and Carter's study, this is now changing, but it is a slow process against the Northern Irish social grain. At the same time intimidation, physical separation, and social divergence have all increased significantly (Darby 1986).

The values inculcated in both school and family preserve a separate cultural consciousness. An opinion survey in 1978 (Krejci and Velimsky 1981: 165) showed that when they have to describe themselves, nearly all Roman Catholics chose Irish whereas nearly all Protestants chose British (or occasionally Ulstermen). These two distinct and separate groups both lay claim to Northern Ireland.

## Political response: real and hypothetical

In such a divided society, where the division is based on the most fundamental of all attributes, the land itself, democracy can only work with special arrangements for the protection of the minority. It cannot work if it is based on relative majorities in a simple territorial division into constituencies, as in the Westminster system. In other pluralistic societies divided into clearly demarcated communities, Canada and Spain for example, the arrangements for parliamentary representation take special account of the needs of minorities. Northern Ireland is an especially difficult case because of the detailed geographical mix of Protestant and Roman Catholic in places like County Armagh, Belfast, and Derry; but still some sort of consociational democracy or 'power-sharing' would be the obvious political arrangement. British incomprehension of the depth of the conflict over the very right to Northern Ireland led to the province being saddled with a political organization based on the assumption that the socio-cultural structure would stand for a straightforward Westminster-style parliamentary democracy. There was no corrective to encourage negotiated settlements between rival groups. This was in spite of the fact that the regional parliament at Stormont, under the single-member constituency method, predictably left Roman Catholics severely under-represented (about one-fifth of the seats for about one-third of the population). Moreover, a property qualification disfranchised one in three of the electorate at local government level, mostly Roman Catholics. Not surprisingly, under all these circumstances, the Roman Catholic minority felt aggrieved.

The British government has been left with the problem of how to deal with the inevitable failure of the parliamentary system in a society where two-thirds of the population claim the right to control Northern Ireland as part of the United Kingdom as a democratic majority, while one-third of the population claims the right to control Northern Ireland as part of the majority in the island of Ireland, from which the six counties of Ulster were artificially cut off in 1921. Unlike the West Bank, there are no wider regional conflicts of which a solution could form part of a comprehensive settlement. Unlike the Falklands, there is no medium-term military solution available. There are two irreconcilable cultural communities whose existence is based on the very right to control the same piece of land.

Successive British governments have tried to find some power-sharing formula acceptable to both communities but it has proved impossible to abandon a system which works so clearly to the benefit of the Protestant majority. The Stormont parliament has been suspended since 1974, but the constituency relative majority system is still in use for Westminster and it has been impossible to persuade the main Protestant parties to participate in any sort of power-sharing arrangement. It is inevitable that British reconciliation, involving as it must some kind of consociation, has been looked upon with suspicion, particularly by the Protestant community which would have to make a political sacrifice. This is even the case if the sacrifice is mainly psychological, as with the 'Anglo-Irish Agreement'.

The Protestant community has been accustomed to being the community with the power, ever since the Plantation. This was reflected in the 1921 division. Ireland was divided in a way that did not correspond to the numerical strength of the two communities. Whereas fewer than a quarter of a million Protestants were assigned to the Irish Free State (later the Republic of Ireland), half a million Roman Catholics were assigned to Northern Ireland.

The districts of Fermanagh, Newry, Mourne, Derry, and Strabane all have Roman Catholic majorities and they are all next to the border with the Republic. Krejci and Velimsky (1981: 187) pointed out that if they all became part of the Republic of Ireland, the respective Protestant and Roman Catholic minorities there and in Northern Ireland would be roughly the same size, offering equal size hostages for some sort of mutual protection. Of course, the same basic problem of the conflicting claims would remain, just the numbers would be more even and the sense of injustice shared. Their further discussion of two purely theoretical solutions, a united Ireland and the unmixing of populations, equally cannot really solve the insoluble problem of conflicting claims but could in the long run take heat from the situation. They suggest a more general model for approaching the problem for conflicting territorial claims, rather than any policy prescription.

The first idea is to start moving formally towards politically uniting the island of Ireland. Instead of a two-to-one Protestant majority in a territory with one and a half million people, it would produce a three-to-one Roman Catholic majority in a territory with four and half million people. At first sight then, the problem of conflict over territory between the two occupying communities would remain with the apparent injustice reversed. But this would not really be so. Unlike the present Roman Catholic minority in Northern Ireland, the Protestant minority in Ireland as a whole would still remain on average more powerful economically in Ulster; it would therefore retain its real control of many important aspects of life in Ulster, assuming that the same economic and political system prevailed as in the Republic of Ireland at present; so it is a moot question whether the Protestants would rebel, although they would certainly be affronted, aggrieved, and fearful.

Krejci and Velimsky's second idea is the unmixing of populations, the geographical separation of people who find it hard to live together. After the Second World War, Germans were displaced from Czechoslovakia and Poland, bringing to an end centuries of conflict; after the Algerian war, many French shifted from Algeria to France despite a strong feeling of a right to be in Algeria where their families had lived for maybe several generations; it was no ordinary colonial settlement, more closely comparable with British involvement in Ireland than, say, India. In Northern Ireland, the unmixing of the population is going on anyway, as a result of violence and the threat of violence, on mixed housing estates and shopping arcades (Darby 1986). It could be encouraged and legitimated without violence and the result of substantial unmixing would probably be easier to police. Still the Protestants of Ulster would claim the right of self-determination in their Ulster homeland and they would still be in the overall majority; still the Roman Catholics of Ulster would claim their right as true Irish to a United Ireland.

The only way in the end for such a conflict over the right to own and control territory to be resolved is if one side wins or loses. That is what happened in Czechoslovakia, Poland, and Algeria. The losers left. One day that may happen in Northern Ireland.

**Power through right: summary table**

*Philosophy of power through right*

| | |
|---|---|
| 1. THE TRADITIONAL NATURE– NURTURE DEBATE | Locke and Rousseau. |
| 2. THE NATURE-NURTURE DICHOTOMY IS WRONG | Examples: children's development, primitive societies. |
| 3. AMERINDIAN PHILOSOPHY AVOIDS THE NATURE– NURTURE DICHOTOMY | Man is at one with the land. |

*A focus for power through right: the territorial state*

1. ECONOMIC  Economic changes meant that territorial control and clear boundaries were important for trade and growth.
2. CULTURAL  The development of the vernacular through printing and the discovery of alternative religions and cultures through exploration brought the great realms of the Middle Ages to an end; the territorial state became the 'natural' form of organization.
3. THUS A GENERAL ATTACHMENT TO THE LAND BECAME A SPECIAL ATTACHMENT TO A TERRITORIAL STATE.

Case-studies relating GEOGRAPHY to POWER THROUGH RIGHT

1. HOLY LAND
   The Jews' attachment to the land has led to the Jewish state of Israel.
2. FALKLANDS
   Two states went to war over claims of superior right to these unimportant islands.
3. GRENADA
   The independent state's right of self-determination is an illusion in an unequal world.
4. NORTHERN IRELAND
   Two irreconcilable claims on behalf of two states to one province.

*The feeling of a right to territory motivates people to fight – sometimes literally – for the land which they believe they should have. This man–land relationship, therefore, profoundly effects the world's political geography.*

# Chapter three

# Power through nationhood

## Introduction

A nation is more than the sum of the people who live in it. The transactions and affinities which define a nation mobilize people by giving them a clear identity. They belong to an integrated group with a shared destiny. But nationhood is so demanding, even going beyond ethnicity in the intensity of its transactions and affinities, that it can only be generated within the legal framework of the territorial state. When the state successfully generates nationhood within its boundaries, the resulting nation-state has much more power and a much clearer identity than a mere formal state. It can do more; its integration makes its people behave more predictably and controllably, which strengthens both the elite's capacity to organize it and its capacity to act within the international political system. The processes by which a territorial state spawns a nation are both geographical and political; these processes of integration are the focus of this chapter.

First, the chapter introduces the philosophy of power through nationhood. The state is weak without the consensus of nationhood, but nationhood needs the organization of the territorial state. The affinities and transactions which make up nationhood are basically the same as those which make up ethnicity, but the sovereignty and territory of the state add a clarity and identity, transforming ethnic connections and similarities into the transactions and affinities of nationhood. This is the process of integration; it results in a nation-state which can claim the loyalty of its citizens, can motivate them, and can mobilize their energy and support.

Next, the chapter looks in detail at the nature of the relationship between nationhood and the state. Various characteristics of the state make it more or less likely that the integration process will take place and nationhood will develop, such as the conduct of the state's elites, certain types of historical changes, military strength, or the significance of subnational ethnicity within the state. But when nationhood and state do succeed in combining, the nation-state is able to take advantage of the immense potential of state sovereignty, its universal, sole, and compulsory jurisdiction within territorial boundaries. The combination of state and nationhood is very powerful indeed.

The purposes of the four case-studies is to see how nationhood can be made and unmade by the state. 'England after the Norman Conquest' (case-study 3.1) provides an example of an embryonic state organization, the result of the imposition of one ethnic group over another, creating an embryonic nation through the imposition of the integrating feudal system.

The scene then shifts to a modern case-study, Guinea in West Africa, chosen because it illustrates the issue of nationhood as a source of power particularly starkly, highlighting the distinction between the state with integration and the nation-state. The African state, created without reference to ethnic boundaries, is not new; 'Pre-colonial nationhood in Guinea' (case-study 3.2) consisted of the Almamy Empire of Fouta Djallon and the Mandinka Empire of Samory Touré, which were integrated into nationhood primarily by Islam in the former case and by a military organization on which the whole empire depended in the latter case. Colonial status, with its exploitation of ethnic divisions and orientation of the economy and society around raw material and cash crop supplies for France, was destructive of the transactions and affinities of nationhood. Independence in 1958 saw specific policies aimed at restoring 'integration and nationhood in Touré's Guinea' (case-study 3.3). Sekou Touré was Guinea's President from independence until 1984. Under Touré, the achievement of nationhood by the functional and spatial integration of Guinea's economy, polity, and society took priority even over economic growth. Under Lansana Conté, Touré's successor, the priorities have been reversed. 'Disintegration, reintegration, and nationhood in Conté's Guinea' (case-study 3.4) examines the tensions created by the new and commoner priorities. It seems that, by the chance location of important mineral resources in remoter parts of the state, some of the transactions and affinities of nationhood will be retained.

Overall, this chapter shows how statehood combines with nationhood to produce a powerful nation-state. The integration of the territorial state is partly a geographical process, since spatial and functional integration are equally necessary for nationhood. Certainly the whole nature of the state, its domestic organization as well as its international behaviour, are fundamentally affected by whether or not the state is really a nation-state. The nation-state is a powerful human unit, whereas a state is only a legal unit. The human geography of a nation-state is therefore very different from the human geography of the legal state. Only a nation-state can effectively mobilize its citizens. This chapter aims to show how and why.

## *The philosophy of power through nationhood*

### Nationhood and consensus

Nationhood is an idea which causes people who believe in it to come together, work together, and fight together. Just like a realization of superior strength (power through might) or a sense of the right to territory (power through right), a feeling of nationhood is a spur to organization and action and therefore a source of power. Nationhood entails a sense of mutual belonging and territorial identity; it implies some degree of mutual understanding and consensus. Talcott Parsons (1967) said that power is the use of authoritative decisions to achieve goals which must be collectively agreed or understood. He and Hannah Arendt (1970) both maintained that compulsion was nearly impossible unless it was backed by a consensus or a negotiated order; Arendt said that tyranny was the most violent and least powerful form of government. The most effective basis for organizing consensus or negotiated order is nationhood.

### Nationhood, ethnicity, and the territorial state

Nationhood is made up of a number of ethnic affinities, principally language, religion, popular ideas about race, mythology and cultural history, association with territory, and political development. Most of these components are both flexible and hard to define, but these affinities are nevertheless crucial in creating a consensus for action and organization. Although they do not need to be associated with identifiable territory, ethnicity has been an important feature facilitating territorial organization through mutual understanding and a sense of mutual belonging.

By the time the era of the territorial state supplanted the great religious and dynastic communities, few of the emerging state boundaries encompassed only one ethnic group. Even allowing for the fact that ethnic groups were very difficult to define, it was rarely the case that one language or one religion or – by most definitions – one racial type clearly prevailed in any one state. The same also applies, often more strikingly, to recently created states. Most are mixed; several ethnic groups, sometimes with one group temporarily dominant and sometimes not, are amalgamated into one state. The state itself has then developed its own distinctive culture and its own history and in this way created a new set of affinities which might also be called 'ethnic'. The territorial state has created a new set of ethnic affinities, based on its defined territory, and it is this state-based ethnicity that may be called 'nationhood'. Intentionally and unintentionally states have carried out their own nation-building, and have thus created 'nation-states'. Both ethnic affinities unconnected with the state and ethnic affinities generated by the state have contributed to a special sense of ethnic identity: 'nationhood'.

Nationhood needs state territory. Every nation is defined by a state, real or imagined, occupied or merely claimed. Although there is a view that an ethnic group needs territory as part of its identity – shared sentiments arising from shared or similar experience (Coulburn 1959), territory is not at the heart of ethnicity. Anthony Smith in *The Ethnic Origins of Nations* takes the view that the essence of ethnicity is contained in name and myth and that whereas it is possible to conceive of an ethnic group without either territory or a claim on territory, there is no ethnic group without a name:

> In general . . . collective names are a sure sign and emblem of ethnic communities, by which they distinguish themselves and summarize their 'essence' to themselves – as if in a name lay the magic of their existence and guarantee of their survival. Like talismans, collective names have taken on mystic connotations of potency; once again, the mythic quality of a name is far more important to the study of ethnicity than any sober account of its origins and practical uses would suggest. A collective name 'evokes' an atmosphere and drama that has power and meaning.
>
> (Smith 1986: 23)

With the name goes the mythology. This can range from an oral tradition about ancestry and origins to a sophisticated series of stories based in part on historical events and in part on the developing nature of society. For example, Hindu myths 'coalesce and are edited into chronicles, epics, and ballads, which combine cognitive maps of the community's history and situation with poetic metaphors of its sense of dignity and identity' (ibid). It is then name and myth and not territory which are the prerequisites for ethnicity, but state territory is a clear prerequisite of nationhood.

## Culture and nationhood

The cultural history of an ethnic group is crucially important for providing it with an unambiguous identity. If the ethnic group is defined by the territorial boundaries of the state, then the impact of cultural development is not just on ethnic consciousness but on national ethnic consciousness and therefore on the development of nationhood.

There are good examples in English, German, and Turkish history. In England, the combination of Norman French and Anglo-Saxon into English coincided with political changes which made the new language a focus for nationhood. In Germany, a feeling that the German language was also capable of being a sacred language played its part in the Northern European Reformation, but the Reformation also retarded wider German national ethnic consciousness as a result of the wars it brought about between German Catholics and German Protestants. Turkish national ethnic consciousness was held back by the commitment of most Turks to Islam, which was of course multinational and based on the Arabic language. Some interesting cases are cited by Sivan (1986: 121–5) in his paper *Languange and Nation: the case of Arabic*; Sivan explains, for instance, that the term for 'socialist' in Arabic, (*ishtirakiyya*), was in fact invented and used originally only by Turks, but when Turkey became secular and culturally separate from the Arab world, *ishtirakiyya* was dropped in favour of the word 'socialist' itself; it was only in keeping with Turkey's newly developed national ethnic consciousness to use a key political word which, while it was still international, distinguished Turkey from the archaic universal Ummah Islam.

## Territory and nationhood

The senses of belonging and solidarity associated with an ethnic group are also dependent on the group's association with territory. There is no case where an ethnic group has been successfully mobilized and motivated to action without a territorial focus; for example, recent moves by the Tibetan leadership in exile to water down its claim for Tibetan territorial independence has had a damaging effect on the morale and cohesion of Tibetans across the world engaged in a peaceful struggle against the Chinese occupation of 'their land' (Thubten Jigme Norbu 1986; Mirsky 1988). When the right to territory (discussed in Chapter Two) is a shared perception across a substantial population, it strengthens ethnic consciousness and contributes effectively to the ability to mobilize an ethnic group. The case of the North American Indian tribes contrasts with the case of the Australian Aborigines (see p. 55); in the Indian case, there was a coherence, in which for example the same landscape features had the same meaning for large numbers of people, whereas in Aborigine groups only very small numbers of people shared the same territorial myths and meanings. But in the Indian case the territories involved were still not viable for the development from ethnic consciousness to nationhood, because there was no concept of a state.

An ethnic group's relationship with its territory must be shared by a group of people large enough to retain some measure of cultural, economic, and therefore political independence and continuity over time. The consequent range of viable sizes for ethnic groups is probably from relatively small tribal units to super-states depending partly on the group's ethnic coherence, but also on its ability to retain its independent identity. The powerful man–land relationship of clearly defined North American Indian tribes collapsed because the Indians could not resist politically, economically, or militarily the encroachment of white America. On the other hand, the Jews in Israel, who had an equally strong ethnic consciousness, quickly became self-sufficient in culture, capital, and labour and were able therefore to be politically independent and retain their ethnic identity; Jewish ethnic consciousness became Jewish nationhood in a clearly defined and independent territory. The significant difference was that the Jews were able to develop a state in a world where the state is the most powerful type of organization; they could develop Jewish nationhood, whereas the Indians only had ethnic consciousness. But even having a state may not be enough; it must also have that degree of political development necessary for a viable ethnic group if it is to be a nation-state with its own nationhood. This helps to explain the difficulty of the task facing many recently formed states in Africa which have attempted a policy of developing nationhood or a national ethnic consciousness. They may be able to promote ethnic affinities artificially within state boundaries with policies specifically aimed at political, cultural, social, and economic integration (Markovitz 1977; Smith 1983), but retaining real political independence is often almost impossible. The final three parts of this chapter examine the case of Guinea in West Africa where nationhood was achieved, against the odds, for a time.

## Nationhood and the state

### The development of nationhood by the state

The state has strengthened and reinforced ethnicity by creating a new shared ethnic consciousness to coincide with the territory covered by the state itself. This is the nation-building process, the creation of nationhood through the promotion of ethnic images and affinities and their application to the state. The sovereign state has indeed proved the ideal forum for the development of this ethnic consciousness – all the various ethnic affinities which together make up nationhood.

Krejci and Velimsky (1981: 3–12) undertook an interesting study of the development of nationhood in modern Western Europe. They identified politicians and journalists as key influential figures in the state and looked at the way they helped to shape ethnic consciousness into nationhood. Politicians on the right have been particularly important in strengthening national awareness. On the left, national consciousness has sometimes been rejected altogether in favour of class consciousness with national consciousness seen as damaging to the cause of revolutionary change. Sometimes politicians have resorted to a national ethnic stance for practical reasons of foreign policy (such as Chamberlain over the annexation of the Sudetenland) or for practical reasons of domestic policy (such as Wilson's appeal to the folk-history of the 'Dunkirk Spirit' at a time of economic difficulty). In Germany, the political use of a nationalist term, '*Volk*', has conveniently shifted in meaning during the twentieth century from 'culturally German' in Wilhelmine Germany through 'racially German' in Nazi Germany to a double meaning in the modern Communist German Democratic Republic of both 'German' and 'Popular' – the *Volkskammer* (Parliament), *Volksarmee* (the Army) and the *Volkspolizei* (Police).

The simple stereotypes of the practising politician are also reflected in political journalism. Four of the staples of the political journalist are internal politics, foreign politics, social conflict, and ethnic feuding. Each of the first three has potential for the promotion of national consciousness, strengthening domestic awareness in the handling of social or political conflict or creating foreign stereotypes. Ethnic feuding itself tends to be treated disdainfully by influential political journalists with an appeal to the nationhood over and above subnational ethnicity. This applies especially to journalists of the right who deprecate, for example, *L'Autonomisme* in France or *Les Contentieux Communautaire* in Belgium or who refer to the conflict in Northern Ireland in religious terms only. Left-wing journalists tend either to deprecate ethnic feuding as reactionary or to identify reactionary and progressive forces without regard to ethnicity.

The general public is influenced strongly by the environment created by politicians and journalists. An atmosphere is created where ethnic national consciousness thrives. Almond and Verba (1963) identified formal education in schools and broader civic education as the most important component of popular ethnic nationhood among the led rather than the leaders. They emphasized the power of education to make acceptable not just ethnic images which coincide with popular feeling but also quite new and artificial national images, significant components of nationhood in twentieth-century creations like Czechoslovakia and Nigeria.

The effectiveness of politicians, journalists, and the world of education in the promotion of nationhood is circumscribed by three main considerations. These are recent history, military strength, and the strength of subnational ethnic consciousness. First, recent history is the simplest ingredient. It is obvious that nationhood is likely to be weaker in recently formed nation-states, like many in Africa. A good European example would be Austria, which combines recent creation with an almost accidental evolution through the breakup of the Austro-Hungarian Empire, the *Anschluss* (unification with Germany), the four-power occupation, and the enforced split with Germany. Again, nationhood can be reinforced in longer-standing nations with a history of foreign conflict such as France and Germany where war has provided xenophobic backing for the development of an ethnic national self-image. Second, military strength is important because the state's ability to defend itself gives it a free hand for the development of ethnic affinities and images; stronger states also provide particularly good opportunities for encouraging feelings of ethnocentrism, a sense of national ethnic superiority, and for promoting negative images of foreigners. Third, strong subnational ethnic consciousness clearly inhibits the development of nationhood. Belgium illustrates this point: the Belgian nationhood is weaker than that of its neighbours despite the endeavours for nearly two centuries of the Belgian state to promote a sense of nationhood, since the country is split between two ethnic groups, the Flemings and the Walloons. After all, the territorial state developed for quite separate economic and cultural reasons, discussed in Chapter Two (see pp. 57–61). Where ethnic consciousness slotted in geographically and historically with the state, then the role of the state in promoting national consciousness, or nationhood, was simple. Of course, the slotting-in was not always accurate, straightforward, or without strife, but the power of the state – Lukes's real power; a capacity, a facility, and an ability (Lukes 1974) – is such that it can usually enforce the consensus of nationhood on its citizens.

## The extraordinary power of the sovereign state

The state, the creator of the nation, is distinct from every other sort of organization. It has universal, compulsory jurisdiction within its territorial boundaries. It has sovereign authority, recognized by other states and by its own people. This emotional meaning is reinforced by the very uniqueness of the state. So much of human life is carried on within it, through it, and by it, that willy-nilly it contributes to the set of ethnic affinities which make up nationhood, incorporating through simple territorial boundaries those who were perhaps excluded by other ethnic groups before the state came into being. The sovereign authority of the state is so strong and effective because it is self-reinforcing; its universal, compulsory, and sovereign jurisdiction reifies the ethnic concept of nationhood, not always painlessly but nearly always effectively.

The state's most obvious unique characteristic is its universal sole jurisdiction within territorial boundaries. Whether or not everyone living within its boundaries is a full or first-class citizen, certain obligations and rights apply to each of them. They all have to pay taxes or dues, they are all entitled to protection under the criminal or civil law, and they all have an obligation to keep within the law. The jurisdiction of the state is applied universally to all people who happen to be within its territorial boundaries – and the jurisdiction of some state cannot be avoided as states divide the population of the world between them. Even those who are called 'stateless' live in some state unless they are literally adrift in the ocean. In the past, such universal jurisdiction might have belonged to the church or to a great empire but now it belongs only to the state.

The jurisdiction of the state is also compulsory. For people living in a state there is no choice about it. If a person lives in, or even visits, a certain territory, he is automatically subject to the jurisdiction of the state which controls it. He may choose to leave for another state but he cannot quit the international system of states. There is no other association which has this compulsory jurisdiction. One may, for example, have no choice but to remain part of a family, a religion, or an ethnic group, but one never has to participate in its affairs or acknowledge it. It may be a sacrifice to leave these other associations, but it is not impossible to escape, as it is to escape the collectivity of states.

The one common function of every state is to settle and prevent conflict; the state sets out to create security within and protection from without. Within the state, law is just one method but order is the principal purpose. Order and predictability of behaviour are achieved in part by law and in part by the social means of inculcating regular modes of behaviour, such as formal education and the media. This security function is basically a negative function. It is protective and preserves the status quo, and it is quite unique to the state. Only the state has as an actual purpose its own internal order. Other organizations do, of course, try to maintain some internal order but with schools, clubs, businesses, trade unions, and so on this is a function subordinate to the main functions of education, profit, or whatever. It is in the fulfilment of this unique function that the state uses its power to inculcate nationhood into its citizens and to become a nation-state.

The negative function of security is unique to the state but positive functions are not. Many organizations have positive functions, including charities, religious communities, and families. All states join these voluntary organizations to some degree by promoting certain kinds of welfare, which may include the basic means of physical well-being (like food, houses, and medical care) or the basic means of spiritual well-being (like education, art galleries, museums, and theatres). The state's involvement in this kind of positive function is also an important part of its negative function; by promoting cultural activity, for example, it is not hard to see how the object of inculcating ideas of nationhood among citizens may be served.

The state provides the rules within which other organizations have to work. They are sovereign rules. The state is sovereign. It is through its sovereign power that the state can enforce both its negative and positive functions; sovereignty gives the state its staying power once the consensus of nationhood is combined with the order and authority of the state and the nation-state has been created. The power of the nation-state created in this way is immense. To say that a nation-state is sovereign is to say that it has supreme and final authority over all its people. The rules of the sovereign nation-state override the rules of any other association within its territory such as a borough, a factory, a company, or a college. All are subject to the authority of the nation-state. They make rules only to the extent that the nation-state requires or permits them to do so, and any dispute between them is subject to jurisdiction by the nation-state's courts. For a state without the consensus of nationhood, sovereignty is much less effective. Nationhood is a motivating force whether or not it coincides with the structure of a state, but the state has the structure to mobilize nationhood – the state gives it power.

Statism: a short-lived phase of history?

This state sovereign power may be only a temporary phase in history. In the past the great religious communities and dynastic realms overrode the territorial state. Now there are cases where state sovereignty plainly gives way to superpower strength, as in Grenada or Poland, and there is also some voluntary submission by sovereign states to the rules of inter-state bodies such as the European Court of Human Rights or the International Monetary Fund. But for most purposes there is still nothing superior to the sovereign state and everything in its territory is subject to its rules.

The system of state sovereign law is very cumbersome. Hobbes (1968) said that people need sovereign authorities to solve disputes or otherwise there would be chaos. But there are many ways of solving disputes such as by debate, mutual consideration, love, respect, and friendship. Occasionally fighting will also play a part in solving a dispute, but states also often use violence both externally and internally. Sometimes it is useful for individuals or groups within the state to apply to the state to act as final arbiter, but earlier simpler arbitration might solve many disputes more quickly. Open disagreement is often healthy and there is always a risk of state-imposed uniformity which may well be worse than having disagreements.

The reason for the attachment which people feel to their state could not possibly lie in its utility; it performs few useful functions. However, because it has clear territorial boundaries and internal sovereign power it is very simple in conception. It controls the people within it by fulfilling a need for territorial organization and by promoting through its organization a set of ethnic affinities which together constitute nationhood. It solidifies both its people's loyalty and its control by the force of its sovereign law and by its power over civic, cultural, and general education. It reifies the ethnic nation and gives people a sense of belonging both to a nation and to a state, in fact to a nation-state. There are signs that the era of the nation-state might be just starting to draw to a close but, for now, bad governments, lost wars, genocide, and economic disaster all bring to an end the lives of particular sovereign states only for new sovereign states to emerge. As yet, there is no real questioning the powerful world system of nation-states and no clear indication of what will succeed it.

*Case-study 3.1: England after the Norman Conquest*

Introduction

England following the Norman Conquest provides a good example of the early development of nationhood. Territorial organization by a dominant ethnic group, the Normans, resulted in a system of transactions and affinities, involving both themselves and the defeated Anglo-Saxons, which amounted to the embryo of an English state and the beginnings of English ethnic nationhood. The Normans were intent on securing themselves and their position in England as soon and as completely as possible, so they set about creating an integrated social and economic system which made conquerors and conquered dependent on each other. They aimed at territorial organization through hierarchical feudal tenure, organized feudal justice, and diplomatic initiatives. When they found themselves unable to create the basis of nationhood in this way, they resorted to violently destructive military campaigns.

The embryonic state generates nationhood through feudalism

At the heart of Norman political organization was the king. The prerogatives and theoretical power of English kingship came from Anglo-Saxon times, but it was the Normans who were able to put theory into practice. For example, the sheriff (who was personally responsible to the king as a local administrator) found his powers greatly increased under the Normans, although the system within which he worked was Anglo-Saxon. Indeed, William the Conqueror's use of Anglo-Saxon institutions made him much more acceptable than would otherwise have been the case; within a couple of years of conquest, most Anglo-Saxon leaders and local militias (the Fyrds) were co-operating with him voluntarily (Douglas 1969: 279).

The Norman invaders quickly became the nominal landlords of large areas; indeed by the time of Domesday Book in 1086 only two Anglo-Saxon landholders of any significance remained (Loyn 1962: 316–17). Nevertheless, the Normans were constantly afraid of retaliation, so it was in their interests to organize the conquered land in the safest way possible, based on mutual co-operation amongst the new landlords and hierarchical co-operation between king and baronage. The invading community instituted a feudal system over and above the looser feudal system of the Anglo-Saxons. A feudal system of government with a monarch at its head was the best available form of territorial organization and the Normans were quick to refine and use it.

Feudal tenure was the organization of society on a war basis; the provision of an adequate military force for the king was its prime aim. William the Conqueror wanted to be able to call upon some 5,000 suitably equipped knights whenever necessary, and the feudal contract between him and his baronage allowed him to do so. The Norman baronage and other tenants-in-chief held lands from the king on the clear and definite understanding that they would provide a certain number of properly equipped knights in return. In Norman England, public service through the provision of soldiers and counsel, personal relationships with the king, and the tenure of land were all combined into a coherent system, much more coherent than anything that had existed in Anglo-Saxon England or in Normandy, in which service arose directly out of contracts and contracts were based securely and permanently on land tenure.

Obviously, real territorial control was very limited indeed in the eleventh century. The central purpose was to secure the role of an elite community, and territorial organization was the best way of doing so. As far as control was concerned, as long as it was understood that the ultimate authority lay with the king, William the Conqueror left his tenants-in-chief free to make their own arrangements for the administration of their estates (Douglas 1969: 273–88). Indeed, there was no serious attempt at uniformity at that level of control until the thirteenth century, when all the tenants-in-chief were English by acculturation and long settled; they were not the Norman Conquest's mixture of Normans, Bretons, and mercenaries from all over Flanders and France.

One of the main reasons why feudal tenure was such an effective way of organizing England, and why it lasted so long, was its flexibility and safeguards. The basic contracts were not rigid but open to interpretation and adaptable to local customs. If the king plainly broke the contract, for example by ignoring hereditary claims, going to war on his subjects, or retaining too many knights in his service for too long, then his tenants-in-chief could express their resentment in a legal and tangible form in a *Diffidatio* (Defiance). The most famous case was *Magna Carta*.

The contractual nature of feudalism and its effect in binding together the king and his tenants-in-chief and eventually, systematically, all tenants with landlords and ultimately with the king in a flexible system of contracts was supported by the feudalization of justice. This was hurried into place by William the Conqueror.

At the peak of the judicial hierarchy was the Royal Court (*Curia Regis*). Its composition was strictly feudal; it was not because the king thought its advice would be useful or because he thought it politically wise to invite them to tender advice that his most senior tenants-in-chief attended the Royal Court. It was because it was their feudal duty to be there, to give service and give counsel. The Royal Court became a court to which his vassals had a right to come and be heard. Laws were made only when the whole court met at the Great Counsels at Christmas, Easter, and Whitsun; they were never made just by the king. This system of centralized law-making and justice was superimposed on a system of seignorial justice of feudal lords in their own local courts and a complex system of Anglo-Saxon law. In this way, the same flexibility was introduced into a feudal system of justice, which became, in effect, a subsidiary part of the feudal system of tenure (Douglas 1969: 284–8).

The feudal arrangements made by the Normans and their fellow conquerors locked them together into a system of mutual support and into a hierarchical system of mutual dependency with the subjugated Anglo-Saxons. It was a system of territorial organization which helped them to feel secure, spread thinly over a wide area, and helped them to retain their power over several centuries due to its flexibility in adapting to changing circumstances. The feudal system was all the stronger for being imposed peacefully in most parts of England. It was a system which did not destroy old unifying factors like Anglo-Saxon law and which cohered the invading community without damaging any but the highest-ranking members of the host community. There were, however, some parts of the country where this organizational approach failed. Then the security methods of the Norman Conquest were different.

### Diplomatic and military subjugation in the embryonic state

The West of England posed the first threat to the invasion after the victory of Hastings by refusing to accept the new regime. William the Conqueror marched into Devon in January 1068 with an army which included many English mercenaries. The local Anglo-Saxon knights and minor landlords in Devon (the local Thegns) stayed on the sidelines and watched. William besieged Exeter for ten days before the terms of the surrender were agreed. Even the woman who had led the resistance in Exeter, Gytha widow of Godwine, was spared and banished (Douglas 1969: 213). William built a motte-and-bailey castle at Exeter to exemplify and symbolize his armed superiority. He won over the Devon Thegns; indeed, later in 1069 they actually played an important part fighting on William's behalf, repelling an Irish invasion.

In this expedition, William also subdued Cornwall, Bristol, and Gloucester with very little use of force, but more through diplomacy backed up with his observable superiority of force, should he have decided to use it. The effective use of the castle was also an important aspect of these campaigns; it was threatening and visibly powerful with its palisades, moats, and stone fortifications in the hands of trained soldiers.

In the West Country there was mainly an English threat to the security of the Norman invaders. In the north there was a significant external threat. There were strong Scottish and Danish forces at large in the North of England, supported by local people. The Norman castles in Yorkshire fell in 1069 and William marched north in the Autumn (Douglas and Greenaway 1953: 150). In these circumstances, brute force was used instead of diplomacy.

As William moved from Nottingham north into Yorkshire, his approach changed and he savagely devastated the land he passed through, sparing no male and leaving nothing behind that could support life. He reached York just before Christmas 1069 and burnt it down. His army now split into smaller bands and systematically destroyed everything in Yorkshire that could be destroyed. The effects were quite obvious from contemporary descriptions of rotting corpses, plagues, and refugees, as well as from the Domesday Book seventeen years later (Douglas 1969: 220–1). The Danes were defeated on the Tees and William crossed the Pennines to defeat

English rebels in Chester and Stafford where, moving south, he was far less brutal.

As he moved back south, William opted again for diplomatic solutions. These were areas where the threat was mostly from local uprisings and not from other kings, and they were areas where the Norman feudal system fitted more easily into the looser Anglo-Saxon system which had been in place before. Norman security would have been reduced by the kind of violence appropriate to Yorkshire if it had been applied to the south.

About a year later, the Danes moved south and allied with a local Thegn called Hereward. They entered Peterborough, burnt and looted the abbey, and massacred many of the civilians. None of this increased their local support, and, of course, it strengthened that of the Normans who bribed the Danes to leave. Thegn Hereward 'The Wake' disappeared into legend (Douglas and Greenaway 1953: 151).

### Conclusion

The Normans combined the adoption of as much as possible of the existing legal and social order with the imposition of a feudal order which knitted together not only the conquering community within itself but also the conquering community with the conquered community. It also bound both communities to defined tracts of land. It was not easy to impose a new order; it might certainly have been easier simply to have imposed a new set of great landlords to replace those killed at Hastings, but the conquerors sought greater permanency and greater security, so they imposed a system in which they had a fixed place. This entailed a system of law and obligation which bound them by contract to the conquered land and to the people who were living on it. The land itself, its extent and contents, became an obsession with the conquerors. By the end of William the Conqueror's reign, all English places and all England's leading inhabitants were recorded in the Domesday Book; many were the subjects of written contracts of service which bound them into a new, tightly organized military, economic, political, and social system run by new masters who had the power to enforce it. Where the threat of force was insufficient, the new order was simply imposed by means of the destruction and obliteration of the old.

Territorial organization was obviously limited and primitive in the eleventh century, but the first two decades of Norman rule were important years in the unplanned evolution of an embryonic English state. There was never any lack of clarity about the land which the Normans set out to claim, conquer, and secure. They introduced and enforced a new set of affinities within and between two ethnic groups that was to have the final effect of creating one predominantly English ethnic group and, thus, English nationhood.

*Case-study 3.2: pre-colonial nationhood in Guinea*

The necessity of integration for nationhood

Integration is a vital ingredient of nationhood. It is a measure of the quantity and quality of transactions among individuals and groups within a nation-state. These transactions can be political (such as participation, chains of command, and the exercise of political power), cultural (such as schooling and cultural exchange), social (such as class relations and intermarriage), or economic (such as trade and transport). All these transactions may be expected to increase in the integrating nation-state and to diminish in the disintegrating nation-state (Deutsch 1979; Soja 1968).

The focus here is on Guinea, a small West African republic with a population of about six million and a territory of about 100,000 square miles (Figures 3.1 and 3.2). Territorial organization in pre-colonial Guinea was formal compared with much of the rest of Africa south of the Sahara. The earliest entity in what is now Guinea that could be called a state, internally integrated and generating its own nationhood, was the Islamic Almamy Empire of Fouta Djallon in the eighteenth century. In the nineteenth century there was a very different sort of pre-colonial state, the Mandinka Empire of Samory Touré, integrated mainly by the military needs arising from its wars with the colonial powers.

Integration: the Almamy Empire of Fouta Djallon

A series of small, ethnically distinct kingdoms in the Fouta Djallon hills adopted Islam in the space of a few years at the start of the eighteenth century ahead of surrounding areas in the south and west in what now constitutes the rest of Guinea and Sierra Leone. Following an Islamic crusade or *Jihad* in 1725, the Fouta Djallon hills became the centre of an empire stretching almost to the border of modern Nigeria; but it was an empire with only loose allegiance beyond the core area of hills themselves, ruled by an oligarchy of political priests (*Ulāma*), mostly from the families of local nobility (Dupuch 1917). Fouta Djallon became an Islamic state late, at a time when Islam was in retreat south of the Sahara and was consequently geographically isolated from the rest of the Islamic world. To survive in these adverse circumstances, there had to be a degree of political, social, and cultural integration in the Fouta Djallon hill kingdoms, promoted by a policy of positive Islamization.

101

First, Islamic Fouta Djallon overcame internal ethnic rivalries and divisions of caste, both of which had been orientated around the abandoned pagan religions. They lingered for some years but were in the end irrelevant to Islamic social organizations. Furthermore, the penetration of Islam in the early eighteenth century acted as a bulwark against the spread of later deviant forms of Islam which, in other parts of Africa south of the Sahara, tended to exploit ethnic, caste, or territorial rivalries in order to gain a foothold. Second, extensive Islamic educational and missionary work helped to secure Fouta Djallon from disintegration by greatly increasing the number of social and cultural transactions within the the core of the Almamy Empire; the *Ulāma* never established a single capital for more than a decade (Labé from 1840 until 1850), but instead moved court regularly and thus established physically as well as culturally close involvement with up to 300 Islamic centres of learning in the Fouta Djallon hills (Hopewell 1958). Third, the integration of the home area of the Almamy Empire of Fouta Djallon made possible the establishment of a superior military organization to gain suzerainty over neighbouring areas, at least in the role of protector of Moslems living or trading there. For nearly a century Fouta Djallon was a powerful empire loosely controlling extensive tracts of West Africa beyond the European outposts – indeed, there were only a few encounters and some limited trade, for example in wild animal products (Diallo 1971, part 3). But European encroachment and the decline of the trans-Saharan trade eventually contributed to the substantial decline of Fouta Djallon by the end of the 1820s.

### Integration: the Mandinka Empire of Samory Touré

The Mandinka Empire of Samory Touré has been idolized in modern Guinean literature for its resistance to colonial powers between 1870 and 1898; titles like *Samory Sanglant et Magnifique* (Traoré 1962) are typical. It is notable for its political and military integration of an area the size of modern Guinea (but extending into modern Sierra Leone and Liberia and not touching the coast), but there was no attempt at the social or cultural integration typical of Fouta Djallon.

First, the Mandinka Empire united politically in one Islamic state some twenty of the tribal kingdoms, mostly of the Malinke people, over which Fouta Djallon had previously established suzerainty; but, unlike Fouta Djallon, the geographical periphery of the Mandinka Empire was important, the suzerain approach could no longer work in the face of the territorial conquests of the colonizing powers. Whereas Fouta Djallon had been satisfied with occasional tributes from the peripheries of its empire, the Mandinka Empire's major military concerns were on its most far-flung territorial peripheries;

Samory Touré had to assert direct control through military governors with whom he was in regular communication, sometimes arranging mass evacuations as part of the scorched earth policy with which he sometimes fought the French (Balési 1976: 18–45).

Second, while close political control was maintained over the whole empire, social and cultural control was weak; the focus of transactions between ruler and ruled was military. The Mandinka Empire was Moslem and even practised a loose version of the Moslem legal code, the *Sharia*, but more traditional values and practices were also permitted – the village *Imam* was in regular political communication with Samory Touré and took political authority away from the village chief, but permitted a variety of Islamic and pagan practices (Person 1969).

The Mandinka Empire was finally defeated by the French in 1898. As a model of military and political integration, it contrasts with the Fouta Djallon model of cultural, social, and political integration.

## Disintegration: the colonial era

The colonial era (1889–1958) was a period of disintegration for Guinea, and Islamic central authority was destroyed. The Islamic leadership of the eighteenth and nineteenth centuries gave way in the 1890s to the colonial administration of French West Africa. Reports from the colonial authorities in the first decade of the twentieth century reveal despair at the weakness of 'chiefs' and the need to build them up into reliable and viable instruments of colonial rule (Crowder 1968a: 165–82). The focus on the chiefs was significant: as heads of clans, they were a counterbalance to Moslem officials and they were divided from each other by disputed territorial boundaries which could be, and frequently were, exploited by the colonial power. In other words, working through chiefs amounted to a 'divide and rule' approach aimed at disintegrating opposition to colonization and had the exact opposite impact to that of Islam on Fouta Djallon and that of political and military organization on the Mandinka Empire.

Apart from the use of chiefs for administration, the French (like other colonists elsewhere) found that development of a small native elite could prove useful for the economic exploitation of French West Africa. The *Ecole William Ponty* and other training establishments in Senegal provided a class of *évolués* in Guinea (Diop 1985). Although there was only a handful of native university graduates in Guinea at the time of independence in 1958, there was a significant elite of favoured technicians, teachers, and others, a class apart, all ethnically Soussou from the area around the colonial capital and main port, Conakry. The class division between these *évolués* and most of the rest of the native population was strengthened by French colonial policy after the Second World War which sought to guide colonies into a closer political involvement with France. The attachment of French citizenship became the overriding aim amongst the *évolué* class.

The economic exploitation of Guinea entailed a colonial economy orientated around the export of cash crops with its legacy of substantial spatial inequality, typified by heavy investment in Conakry and neglect of all but a handful of main towns elsewhere. For most of the colonial period, the planting and export of coffee and bananas and associated railway development constituted the main economic activity. Despite a gold rush in 1900, it was not until the 1950s that gold, along with bauxite and diamonds, was mined in significant quantities. The consistent pattern was of primary products mined or grown for export, a transport system entirely operating for this purpose, and the whole range of this activity run by foreign companies using Soussou labour from the Conakry area. The export economy was a few miles either side of the railways from Conakry to Kankan and Conakry to Fria. The rest was essentially a subsistence economy with apparently less inter-trading than in the precolonial period (Goerg 1986).

In contrast with much of French West Africa, Guinea also developed an organized working class, albeit on a very small scale consistent with the amount of industrial development. Its organization originated with small groups of *anciens combattants* returning from the First World War, influenced by French socialists and by other aspects of their war experience (Crowder 1968b; Summers and Johnson 1978). *Anciens combattants* were heavily involved with strikes in 1918 and 1920 in the docks at Conakry, and they established there the roots of the only trade union movement in French West Africa to provide a post-colonial leader – Sekou Touré went from the leadership of the *Confédération Nationale des Travailleurs Guinéens* into the leadership of Guinean independence politics. The political importance of the Guinean trade union movement allowed Guinea to escape political leadership dominated by *évolués*, and thus to reduce the influence of *négritude* philosophy, which tends to isolate African culture from politics; Touré and his fellow trade union leaders were influenced by the revolutionary conclusions of Fanon (1969), Nkrumah (1964), and Cabral (1973). Their alternative to *négritude* philosophy was the psychological reconstruction of the African to a revolutionary consciousness, the devolution of maximum political and cultural power to the village (villagers not chiefs), and authoritative democratic criticism and self-criticism of the national political centre (Touré 1959; Jinadu 1978).

This political philosophy helped to bring about the circumstances of early independence for both Ghana and Guinea. The Guinean nation-state had been defined in terms of its political extent in a treaty between the British and French in 1904, but was split economically, officially, and culturally, as well as territorially, at the local level by clan chiefs. Independence in 1958 marked the start of a period of much-needed reintegration.

*Case-study 3.3: integration and nationhood in Touré's Guinea*

A decision to integrate

The leader of Guinea's independence movement, Sekou Touré, took the view that Guinea's continued participation in international capitalism would lead to a continuation of the disintegration experienced under colonial rule. Although Guinea would have political sovereignty, the loss of economic sovereignty would force Guinea to make any other aims secondary to conventionally measured economic growth, the maximum output from cash crops and mineral deposits. Touré's government opted for policies consciously and specifically aimed at the establishment of Guinean nationhood through integration of the Guinean state, politically, culturally, socially, and economically. This was to be at the expense, where necessary, of economic growth.

Integration: external stimuli

Measures aimed at the integration of the Guinean nation-state were assisted by international events over which Guinea itself had little or no control. Guinea rejected in a referendum the opportunity for semi-independence in the French African Community, and the French reaction was violent. The French press and de Gaulle accused Guinea and Sekou Touré of being dominated by the Soviet Union (Panaf 1978: 11–16). All French personnel and assistance was immediately withdrawn, and they burnt the files (Adamolekun 1976: 80–1). The *évolué* elite, suspicious at first of Sekou Touré's independence from the French African Community, drew their own conclusions.

At the same time, Guinea took on all the paraphernalia of an independent nation-state: a flag, an anthem, and laws and speeches referring to 'liberty', 'dignity', and 'sovereignty'. While these all helped to establish Guinean identity within established territorial boundaries, they meant little to a disintegrated and divided state until it was faced with four significant external threats.

First, in 1960, there was the French Army Dissidents' Plot to

*Figure 3.1* Guinea

*Figure 3.2* Guinea's main centres

regain Guinea for the French African Community; the general resentment at the interference from the former colonial power brought to an abrupt end some increasingly violent inter-tribal fighting between the Soussou and the Foulah. The Soviet involvement in the Teachers' Plot of 1961 had a similar effect with the Malinke and the Foulah, and it resulted in the expulsion of the Soviet ambassador. The consequent friendship with the United States was short lived when a diplomatic incident in 1966, involving Guinean ministers arrested on an American plane in Ghana, brought about riots in Conakry and the arrest of American diplomats and Peace Corps workers.

The fourth and most serious threat to Guinea, and the single event which helped more than any other to solidify Guinea's national self-image, was the Portuguese-led invasion from the Portuguese colony of Guinea-Bissau in November 1970. This invasion was intended to release Portuguese prisoners-of-war held by the Guinea-Bissau and Cape Verde independence movements, which it did, and to bring about the overthrow of the Sekou Touré regime, which it did not. The resulting open battle was the first time the mass of the population had been faced with the need to take up arms to defend the Guinean nation-state or, rather, with the choice as to whether or not to do so. That it did fight is to the credit of the integrating nation-building processes that had been going on in the twelve years after independence.

Ethnic integration

One of the Touré government's most urgent tasks had been ethnic integration in Guinea. The French colonial authority's use of ethnic and tribal disputes to promote the power of the chiefs had been a major force for disintegration. A typical example of ethnic division was the inability of Foulah, Guerzé, Malinke, and Soussou to agree on locations for any of the three secondary schools promised by the French for five years prior to independence, when there was local control of such matters. When Sekou Touré's *Parti Démocratique de Guinée* took power under the French shortly before independence, the status of chieftaincy was abolished by law; this was effectively the starting-point for a succession of policies aimed specifically at ethnic integration.

From independence, the ruling party played ethnic arithmetic, sending members from one ethnic group to run the party in an area dominated by another ethnic group, emulating Sekou Touré, a Malinke, who had been mayor of Soussou Conakry. A Soussou minister was given responsibility for Foulah affairs and key posts were generally distributed evenly. Special steps were taken to ensure that the *Assemblée Nationale* contained ethnic representation in accordance with the estimated population of each ethnic group. This also incidentally helped to consolidate support for the *Parti Démocratique de Guinée*. The political patronage that arose from the one-party state which soon emerged was distributed through open positive discrimination to avoid domination by better-educated Soussou from the capital (Ministry of Education and Culture under the auspices of the Guinean National Commission for UNESCO 1979; Rivière 1971).

Because ethnic groups in Guinea were concentrated in certain geographical areas, Sekou Touré also embarked, as part of the general aim of ethnic integration, on ethnicity-based spatial policies. Ethnic equality was given such a high priority that national economic development was frequently sacrificed intentionally in favour of ethnicity-based regional equality. For example, bauxite mining with substantial foreign policy earnings was slowed to 60 per cent of output before independence in the 1959 plan in the Foulah areas east of Tougué where it had been developed by the French, until iron ore production could be developed in the Kissi Forest Region in the 1964 plan following discoveries by Alusuisse. Another example is the pricing policy for rice which favoured inland Malinke production to equalize its price at Conakry, despite being some 20 per cent more expensive in real terms because of transport costs: the policy lasted from 1960 until 1985.

Colonial development, which had favoured the capital and railway towns, was reversed wherever possible, despite high production costs resulting from inferior infrastructure in remoter locations. The *Plan Septennal 1964–71* was intended to put into effect regional sub-plans which were audaciously biased against export-orientation and in favour of spatial equality. The plan anticipated a very slow overall increase (some 0.8 per cent per year) in the standard of living during the plan period. It was a conscious integration exercise. The official report (Ministère du Domain Economique 1970) showed an overall 20 per cent underachievement after five years, so the regional emphasis of the plan meant an actual decline in average living standards in Conakry and four other main towns.

### Class integration

Class integration policies complemented ethnic integration of policies. For class integration, the integrative role of Islam in Fouta Djallon and of the military organization of the Mandinka Empire was copied by the *Parti Démocratique de Guinée* in Sekou Touré's Guinea.

The party became completely dominant within a year of independence (Adamolekun 1976: 77–87). No human activity was outside the scope of party involvement from baptism to the channels of food supply, from primary education to funerals. Everything was done 'for Party and Nation'. Nearly everyone in Guinea was a member of the party and had compulsorily to attend party indoctrination meetings.

The party youth organization (*Jeunesse de la Révolution Africaine*) incorporating all Guinean youth, was set up in 1959 and replaced, where there had been any organization at all, a variety of regional youth groups dominated by the chiefs. The focus of the activities of the *Jeunesse* was Guinea as a whole – national and party cultural, sport, national defence, and revolutionary thought. This did not mean that ethnic origins were overlooked; it would not have been possible to effect any sort of real cultural revival in Guinea without reference to ethnic culture. Youth organizations had used French as their *lingua franca* in the colonial period; now they used all three main national languages and five minor ones, which were also used as teaching languages up to the second year of secondary schooling; a large *Académie des Langues Nationales* translated numerous texts from French into all eight languages. The *Jeunesse* thus had the twin aims of reducing the influence of residual *évolué* francophilia and of counterbalancing the ethnic class distinctions promoted by the French who had incorporated virtually only the Soussou from the coast in the trading economy, except for limited amounts of heavy or menial labour.

The objective of overcoming class distinctions by using the party machinery could also be seen in the rules governing party membership and especially membership of the party elite. The party was intended to be an agent of social equalization, promoting the labour of the industrial workers in the towns and the peasantry of the villages at the expense of capitalists and the elites of the colonial period such as the families of the chiefs. Indeed, at various times, the party excluded tradesmen, businessmen, and industrialists from any positions of rank. The party elite themselves, with few exceptions, did four months a year compulsory rural labour service. No new party elite or cadre class endured, partly due to these sorts of measures and partly due to the territorial decentralization of power – and the consequent reduction of power of the central party cadre class – associated with the 1973–8 plan.

During the 1970s widespread powers to determine local policy were given to the villages. Indoctrination had penetrated deeply enough for genuine devolution to take place, along with real popular participation, without risk to the central government of Sekou Touré and the party. Weekly *Assemblées Générales*, each with a part-time executive *Comité de Village*, ran the economic and judicial affairs of the village. The central government still retained control of funding major projects, but the *Comités* and *Assemblées* could decide between, say, cash crops and subsistence crops, roads and school buildings, collective and hierarchical forms of organization for workshops and farms, and so on. This was a new departure for Guinea which had concentrated all but the most minor decision-making in Conakry both under the French and during the first fourteen years of the Touré regime. It is significant that power in Guinea was either at national or village level. Unlike many African states, there was no ethnic regional administration in Guinea. Touré favoured this approach on the grounds that regional administration was as remote as national administration but a focus for ethnic rivalry (Touré 1969: 85–115).

None of these policies, aimed at the ethnic or class integration of the Guinean nation-state, was designed specifically to promote conventional economic development. The argument was that the majority of Guineans had limited economic expectations, at least in the short term, perhaps just food, clothing, and shelter, all of which could be provided at reasonable standards by the Touré regime, in part due to the happy accident of extensive mineral deposits. Touré argued that the cultural assertion of the African and the construction of nationhood in Guinea depended on a measure of isolation from international capitalism, which would inevitably bring with it a new emphasis on economic development over political and cultural development. Since this policy could not be carried out together with other African states, as both he and Nkrumah of Ghana had hoped, it was necessary to isolate Guinea alone. Openness, he argued, would bring about the disintegration of Guinea (Touré 1959, 1969). Since the military coup which took place three days after Touré's burial in 1984, the disintegration hypothesis has been put to the test, and all the initial indications are that it is valid.

*Case-study 3.4: disintegration, reintegration, and nationhood in Conté's Guinea*

Disintegration: the first phase of the Conté era

Three days after Sekou Touré's burial, a military coup (led by one of Touré's military chiefs, Brigadier-General Lansana Conté) took power. Conté immediately announced a programme of economic liberalization to be piloted by a new ruling *Comité Militaire de Redressement National* over which he would preside. Touré's policies for national integration were to be replaced by policies giving absolute priority to economic growth. Within a few weeks the

*Parti Démocratique de Guinée* was disbanded along with the youth organization and its associated militia. Two months later, village organizations were also disbanded, pending 'enquiries into how they could be more efficient'. In a succession of announcements, the new regime made it clear that there was now to be a clear priority for economic development over national integration, and private foreign and domestic capital investment was to be welcomed in virtually every case.

In the first two years of this new regime, the rate of capital investment increased by 114 per cent. Of total new investment, 56 per cent was in Conakry and the surrounding area, 31 per cent outside Conakry but in bauxite mining or ancillary activities, and only 13 per cent (an actual reduction of about 25 per cent over the last two years of the Touré regime) in other activities in other places. Agricultural investment fell back by some 75 per cent under the new regime, whereas investment in non-bauxite mining activities, particularly gold and diamonds, increased by some 120 per cent (Cheveau-Loquay 1987). International aid for small indigenous enterprises has been used almost exclusively for mining ancillary industries, but these are rarely locally owned – for example, when there was an application by a local mineral prospecting consortium in Boffa for enterprise funds, they found the contract for their area had been handed over without consultation to Arédor, a mining multinational, which had agreed to set up a locally run subsidiary (Slowe 1988).

The population in Conakry grew from 300,000 to about 750,000 between 1984 and 1987 with a series of big shanty settlements and appalling urban conditions. Guinea's ethnic disintegration is reflected in the capital in various ways. For example, the dismissal of 10,000 civil servants, under agreement with the International Monetary Fund (IMF), was carried out mainly along ethnic lines by the Soussou-dominated military government. Malinke people were over-represented five times in the sackings with the obvious result that the new poor were nearly all non-Soussou, and the sharp cuts in agricultural investment left them little alternative but to exist on the fringes of urban life. Again, the heavy concentration of both foreign and domestic investment in the capital (with the main port and only international airport), combined with underinvestment in agricultural villages, led to a sharp increase in the urban population by job-seekers from non-Soussou rural areas (Bureau Economique du CMRN 1986, 1987).

The statistical evidence is scanty, especially for comparisons with the Touré era, but all the evidence is that a process of national disintegration was underway along ethnic, class, and spatial lines, as

Guinea started to play a full part in international capitalism. If trends had continued, only the formal state would have remained without nationhood. Yet economic developments in the 1987–8 period have shown promise in some areas of sufficient advantages from participation in the international economy to overcome problems associated with disintegration. It may yet be that the trend to disintegration has stopped.

Overall, whereas Touré's regime targeted ethnic and geographical integration, Conté's regime has aimed primarily at economic development. Touré consciously sacrificed overall economic achievement in favour of geographical, class, and ethnic integration, but the Soussou have been the clear political and economic beneficiaries of Conté's regime. Conté has risked, in particular, the wrath of the Malinke people of Guinea's interior in favour of his own Soussou people based around the capital city, Conakry.

Conté's regime has generally gone out of its way to make itself attractive to international investors. Its reward has been IMF-related investment (on IMF terms) and a close relationship with France. But Guinea's full participation in the international system may yet fail, as Conté himself has recognized: 'We do indeed have a lot of problems here. The investors who come here aren't serious investors. They are more like invaders . . . . Negotiations rarely end with anything positive' (Conté 1988: 23).

Reintegration: the second phase of the Conté era?

There are some indications from political and economic events in Guinea in 1987 and 1988 that the unplanned geographical dispersal of Guinean economic development may overcome some of the disintegrative effects of the Conté regime's policies. It may be that the welfare benefits of integration may be starting to coincide with the benefits of accelerated economic development as the post-Touré era enters its fifth year. First, political reintegration is considered.

Political events in 1987 and early 1988 were the culmination of growing ethnic tension and dissatisfaction with the economic consequences of Guinea's IMF package with its emphasis on major cuts in the civil service. Additionally, among intellectuals, there was a growing frustration at the Conté government's failure to grant any political reforms.

Recognizing the dangers of mounting ethnic tension, Conté undertook a major programme in August and September 1987 to win over Malinke support for his regime with the aim of including more 'trustworthy' Malinkes in the higher levels of the army and government. His failure to obtain this political support produced an angry speech in October in the Soussou language to a Conakry audience in which he referred to an illegal opposition movement, the *Union Mandingo*, which was based in Malinke areas and would have to be wiped out. At the same time, Conté increased his support among the Soussou by promising them greater trade union freedom.

In December 1987, increased Malinke opposition forced Conté to cancel a state visit to France (Andramirado 1987). Then, in early January 1988, he came under further pressure. Students rioted at the University of Conakry, ostensibly against their own specific problems of food and living costs. They were soon supported by Soussou and Malinke workers in Conakry protesting about general economic conditions, especially the price of bread, rice, and petrol (and consequently transport) which all had doubled in the twelve months to January 1988, as subsidies were phased out as part of the IMF package. The riot was put down with one dead and ten injured; but the next day Conté climbed down. He did so under cover of an attack on economic saboteurs and unscrupulous merchants. The government announced price freezes (except on petrol) and improvements in students' living conditions. Having achieved a truce, the regime then went on the offensive. The university's rector, Aboubacar Somparé, suggested there were political forces, controlled by Guinean exiles in the Ivory Coast, behind the students, and rumours of a plot in the making were given credence. The result was a purge of senior posts with new government positions being filled by civilian and military figures considered personally loyal by Conté (*Africa Events* 1988; Barry 1988).

President Conté walks a political tightrope in two major respects. The first is that his dependence on external financial support has forced him to accept an IMF package which includes phasing out subsidies, devaluing the currency, wholesale sackings in the public service, and a concentration on investment for export; this militates against political stability by making it impossible to give priority, as Touré's regime had done, to ethnic equality and national integration. Second, continuing economic liberalization means a loss of wealth and power for the urban middle class who are part of, or do business directly with, government organizations; yet it is they who are inevitably responsible for putting that liberalization into effect. Conté has called for greater administrative efficiency and greater integrity in carrying out reforms. It is an open question how long President Conté can survive the contradictions his regime has created.

While the political consequences may be uncertain, Conté's economic policy is not. It is dominated by the need to cut state subsidies to industry, agriculture, and the civil service and to achieve an appropriately valued currency to allow Guinea to operate and participate more easily in the international economy. Guinea's three-year development plan, initiated in 1986, is concerned with phasing out subsidies, promoting privatization, developing economic infrastructures, and controlling the exploitation of Guinea's immense mineral wealth (Bureau Economique du CMRN 1987; Traoré 1988a).

Privatization is deemed an urgent priority by the Conté regime. Badly managed public concerns are considered a burden on the Guinean economy. Some have already been put into private hands, with a preference for Guineans wherever they have the requisite capital and skill. Two agro-industrial units (producing tea and quinine), a printing works, a brickworks, and a fruit juice factory have all been taken over by Guinean nationals in the last year. But they are not the biggest firms. In these cases, the government tends to consider foreign investors, who almost all complain about the government's tardiness in its gradual revision of Touré's laws in favour of indigenization and against foreign ownership. The government in turn considers that foreign offers are too low and there is now a series of prolonged negotiations going on.

For example, the Sanoyah textile mill, the pride of Guinea's state companies, is the subject of ongoing negotiations between the government and the UCO/Schefer Group. The factory has been at a standstill since September 1987; the workers are still being paid, but the machinery will rust if the situation goes on too long. Again, the Guinea National Hydrocarbons Board, another big concern which could be privatized, should soon be taken over by a consortium of Shell, Total, and Agip. Negotiations have been dragging for two years and the state is still paying the wage bill (Traoré 1988a: 19–21).

The government is very disappointed by the poor take-up by multinational capital of its privatization programme. It now feels that the economic infrastructure has been sufficiently improved with better roads, good water supply, simplified tax systems, and mining and investment codes. It is, therefore, only policies which will assist the broader development of the economy, improving infrastructure in rural as well as urban areas – in other words, policies for national integration that will make the privatization programme work. This has been the focus of substantial development assistance resulting from Guinea's agreement with the IMF.

With FF600 million, Guinea in 1987 was second to the Ivory Coast as a major recipient of French aid. Japan granted Guinea a loan of 5.5 billion yen on IDA terms and 600 million yen in grants as part of the special IMF Facility for Africa. The IMF-World Bank programme is for $115 million p.a. for three years, two-thirds of it to finance projects and a third to help the balance of payments.

Under Lome III, the EEC committed in February 1988 a 70 million ECU programme centred on Guinée-Maritime and Haute-Guinée. All these loans may eventually be converted to grants or longer-term interest arrangements as a result of the decisions of the 1988 Toronto Economic Summit (Marchés Tropicaux 1987a, 1987b; Riegel 1988).

In rural development, some £300 million will be used over the period 1988–91 to develop health care and roads. As far as health care is concerned, there is a major priority programme for the rehabilitation of health infrastructure. As far as roads are concerned, an emergency programme gives priority to a huge primary network rehabilitation scheme. The idea is to asphalt 1,200 km of road, mainly the Conakry–Kankan highway which will be extended with the Mamou–Faranah–Kissidougou–Nzérékoré stretch, and also the Conakry–Forécariah road will be improved (Figure 3.2) and there are plans for 1,900 km of rural tracks as well. The whole programme should mean a dense, properly constructed primary and secondary network by 1993, making for easy access to production areas and removing a major barrier to Guinea's mineral and cash crop exploitation potential. The production of rice, coffee, oil-palm, mangoes, pineapples, and bananas is expanding quickly, but it is mineral exploitation where infrastructural aid has had its most significant effect (Marchés Tropicaux 1987c; Riegel 1988).

Guinea has immense mineral wealth. It has the purest bauxite in the world, producing (under the auspices of a multinational consortium headed by Alusuisse) about 14 million tonnes a year from three sites. At this rate there are known to be reserves for 500 years! There are 6.5 billion tonnes of iron ore reserves, including very high-grade (70 per cent plus) ore from Mount Nimba. Ore exports via Liberia are now planned to start at 6 million tonnes in 1990, the high quality making exploitation profitable even at low world prices. Guinea also produces diamonds; the purity of the stones means demand from the jewellery industry is heavy. The estimated reserves are 400 million carats, two-thirds of them of gem quality; the principal new capital for diamond exploitation is coming from the multinational, Arédor. Guinean gold is historically recorded as the basis of the region's wealth from the fifteenth to eighteenth centuries and gold is still mined mostly by craft gold workers operating individually or in small co-operatives; Union Minière, a French-based consortium is now starting to organize the industry. There is also evidence of offshore oil, for which the major international oil companies are now prospecting (Marchés Tropicaux 1987d; Traoré 1988b).

Guinea's mineral wealth is no longer regarded as the basis for self-sufficiency politically and culturally as well as economically.

Instead, Guinea has now opted for a path of maximum resource exploitation firmly rooted in international capitalism. It may well be that Guinea's prodigious mineral wealth enables her to survive in such a world even when credit and foreign markets are less accessible than now.

The years 1987 and 1988 were politically threatening for the Conté regime. Above all, its administration of scarce resources and its handling of the civil service and public sector cuts, ever since 1984, have been ethnically divisive. On the other hand, the more rapid recent development of Guinea's natural resources has handed to the Conté regime an opportunity to rectify some aspects of ethnic and class disintegration resulting from its policies. International insistence on infrastructural aid and international capital's overwhelming concern with Guinea's geographically dispersed mineral wealth and cash crop potential may now be forcing the Conté regime to recover some of the ethnic and class integration achieved under Touré, and at the higher economic levels of attainment possible in the light of continuing economic development.

It is clear that integration contributes to the development and preservation of nationhood, and consequently to the stability of the nation-state. The case of Guinea highlights the difficulty of pursuing integration and economic growth simultaneously. The pressure from the outside world is always to pursue economic growth, and this pressure was eventually successfully exerted on Guinea.

Sekou Touré was fortunate in having pre-colonial integration models to use and mineral wealth to fall back on, but after his death the outside world became irresistible. When Guinea opened up its economy to the world after the death of Sekou Touré, it risked disintegration politically, culturally, socially, and economically. Its statehood was never at risk, but the affinities and transactions that integrate the state and make it a nation-state were threatened. The next few years will tell whether Guinean nationhood survives.

## Power through nationhood: summary table

*Philosophy of power through nationhood*

1. Transactions and affinities *without* the framework of the state make ETHNICITY.
2. Similar transactions and affinities *within* the framework of the state lead to NATIONHOOD (These transactions and affinities may be called 'integration').
3. Nationhood provides the state with consensus and with the power to *mobilize* and *motivate* its citizens and to obtain their

loyalty. This is the powerful NATION-STATE.

Notes: (a) Certain state characteristics have a bearing on whether or not the state is likely to generate nationhood and so become a nation-state, in particular ethnic groups within the state can arrest the development of nationhood if they are strong.

(b) Certain *legal* state characteristics make a combination with nationhood characteristics even more formidably powerful; these are the characteristics of *sovereignty* – the state has universal, sole, and compulsory jurisdiction over territory.

Case-studies relating GEOGRAPHY to POWER THROUGH NATIONHOOD

1. ENGLAND AFTER THE NORMAN CONQUEST
An early state becomes a nation-state – integration through feudalism.
2. PRE-COLONIAL NATIONHOOD IN GUINEA
Multi-tribal states integrated by religion or armed force.
3. INTEGRATION AND NATIONHOOD IN TOURE'S GUINEA
Economy and democracy sacrificed for integration and nationhood.
4. DISINTEGRATION, REINTEGRATION, AND NATIONHOOD IN CONTE'S GUINEA
Will Guinean nationhood survive?

*The nation-state is a much more powerful entity than the aspatial ethnic group or the legal entity of the state. It is the most consistently powerful organization in the political geography of the world today.*

Chapter four

# Power through legality

## Introduction

The legal distribution of power is the focus for the majority of wars between states. It may well not be the root cause, or even a particularly important cause, but it is the focus. State boundaries, the most potent legal or administrative facts of political geography, directly affect the lives and welfare of millions around the world; when wars are fought because of state boundaries, then the millions may rise to billions. Different kinds of boundaries have different potency and therefore different effects on people's lives; the effects of state boundaries are the main outcome of 'power through legality'.

The first part of this chapter introduces the philosophy of power through legality which explains the links between fundamental values and conflict over legal issues. The most important source for this is the Tagil model, developed at Lund University with an input also from political conflict theory. Tagil provides a model of boundary conflicts which, first, explains how the dominant basic values of the state, modified by prevailing structural conditions and specific current events, lead to the selection of a 'primary' objective. The second part of the model explains how the same structural conditions and events modify further the 'primary' objective into a more limited 'situational' objective which, in turn, leads to action (subject to various negative 'barriers' and positive 'resources'). The results of the action lead to a revised 'situational' objective. The model is calibrated by political conflict theory which identifies the triggers needed to put the Tagil model into action.

The second part of the chapter links the ideas in the first part to the later case-studies by examining the nature of state boundaries. Their delimitation and administration determine the effect they have on the states they enclose and on the people who live near them or want to cross them. The boundary is a barrier, sometimes porous and indistinct but sometimes as solid as the Berlin Wall.

The case-studies reflect four main issues of power through legality. The first case-study is the case of the Sino–Soviet boundary dispute; this is a particularly clear example of a dispute arising from a conflict of dominant values surfacing as pitched battles over boundary lines. The second case-study is of a boundary where there is no clash of values, which is used as a barrier of convenience; this is the case with a federal boundary (Britain–Channel Islands) which is deftly used to the financial, administrative and political convenience of either side. The third case-study, Western Sahara, illustrates a struggle to create a new state within former colonial boundaries; nationhood may eventually follow but, for now, the aim is simply to create a new legal entity, a sovereign state with accepted international boundaries. The final case-study deals with the consequence of a boundary that has disappeared, the end of the partition of Jerusalem. Jerusalem has, for over twenty years, been a mixed city shared between Israeli Jews and Palestinian Arabs, but for eighteen years prior to this, it was partitioned between two states; the scars of partition have been desperately hard to eradicate.

Overall, this chapter shows the importance in world politics of the boundaries that legally define the territorial state. The politics of the boundary influence the geography of the state as a whole and its borderlands in particular, and the geography of the boundary affects the politics of the state and its neighbours. This chapter aims to examine this important two-way relationship over boundaries and its impact on the distribution of power.

## The philosophy of power through legality

### Man's need for legal definitions

> Man needs to find and impose order in the world; to tell differences; to see patterns; to discriminate figures and objects; to make clear distinctions. Man seems to possess a genetic disposition to think in binary oppositions.
>
> (Strassoldo 1982: 246)

The development of civilization, like the development of individuals, is a process of increasing refinement in distinguishing and discriminating between objects, in identifying new distinctions, and in drawing new boundaries. But it is also a drive towards harmony and mutual understanding.

There are two sides to man. There is this drive for defining opposites, for codification and organization, and there is also his alternative drive for harmony. There is the drive to differentiate between people, functions, and places and to draw boundary lines between them, and the desire of the artistic and the sensitive for harmony and similarity and the longing for the destruction of boundaries between people. The lawyers and the administrators of the modern state distil man's awareness of his power to conquer, his rights of ownership, and the group to which he belongs (power through might, right, and nationhood) into objective and measurable pieces of territory; they codify it and set it in tablets of stone and, in doing so, fulfil a need in man.

### The Tagil model: dominant values and primary objectives

Tagil et al. (1977) provide a useful model of the way in which value conflicts over might, right, and nationhood come to be expressed in practice in actual boundary conflicts. In the Tagil model, there is always some particular dominant value of might, right, or nationhood which provides an initial trigger; a trace may then be made, using the Tagil model, from one participant's (or 'actor's') values to his victory or defeat in an actual conflict focused on a boundary issue.

The starting-point is, then, the dominant value which could be categorized under the heading of might, right, or nationhood – or some combination of these. The dominant value might be a combination of ethnicity (nationhood) and economic assertion (might). This may cause the actor (maybe a state, maybe an ethnic group, maybe some other group) to attach special importance to an area with major natural resources thought to be needed for the alleviation of poverty or to be particularly significant for the profitability of major private companies. At the same time, a nation-state or an ethnic group, which perhaps used to occupy the area in question, may be perceived as having been unjustly constrained or poorly treated in recent history and deprived of its potential wealth and glory.

The dominant value in turn plays a leading part in determining the 'primary objective' of any particular actor. The primary objective might be, for example, the securing of some specific territory which contains useful natural resources. The primary objective, however, is not exclusively dependent on the dominant value; its exact nature is dependent on several other factors which Tagil calls 'events' and 'structural conditions'. Events might be historical occurrences, such as subversive acts in the disputed territory which arouse suspicion in the minds of one of the actors. Structural conditions are basic given facts, for example the size, military strength, or diplomatic significance of the actors in the dispute. Because of the intervention of events and structural conditions in the determination of the primary objective, it may well turn out that similar dominant values could lead to different primary objectives.

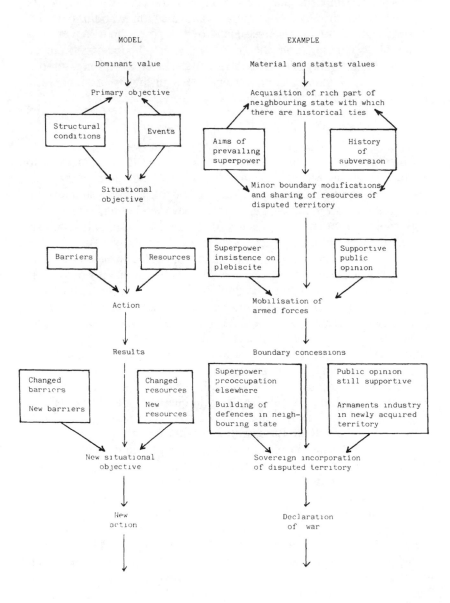

*Figure 4.1* The Tagil model

The Tagil model: situational objectives, action, and actors

New or changing events and structural conditions may add a further dose of reality to the primary objective, persuading the actor to redefine his objectives to meet a new situation. For example, prevailing regional superpower strength or the aims of the victorious states at a peace conference after a war may cause an actor to refine his primary objective into a 'situational objective'.

To get from objectives to action is the purpose of the next stage of the Tagil model. Clearly, the main motivation for a particular choice of action is the situational objective; but there are two other determinants as well, which Tagil labels 'barriers' and 'resources'. Whereas the situational objective is based on the actor's subjective values and his subjective perception of reality, barriers and resources are accepted by all actors as objective realities. A barrier, for instance, might be the public and inflexible insistence by a militarily stronger superpower on self-determination by plebiscite in a disputed area. A resource may be, for example, a state of public or parliamentary opinion, open to subjective interpretation but nevertheless a crucial objective factor in determining actual political and economic decisions.

The situational objective, modified by barriers and resources, leads to action, such as the decision to mobilize armed forces or the decision to hold a plebiscite. This, in turn, leads on to favourable, unfavourable, anticipated, or unanticipated results; these are appraised and a further revised situational objective worked out. By this time, new barriers or resources may have become influential and so a new strategy may be adopted. This time the new strategy and new actions may lead to out-and-out armed or diplomatic conflicts with a neighbouring state. The boundaries of the state, the legal limits of state sovereignty, may eventually be altered, depending on the success of one actor's strategy against another.

Who are these actors in a boundary dispute likely to be? In most of the examples given in this chapter, they are the governments or would-be governments of states, but of course there are many subsidiary actors who may be involved in a boundary conflict. These could typically include the representatives of minority groups living on either side of the boundary in question, political parties, parliaments and local or regional authorities, the state administrative apparatus (including boundary commissions, conference delegations, committees of inquiry, and so on), or the military. There may also be international actors in a boundary dispute, such as multinational business interests, various religious groups and international organizations which can range from the specifically political and territorial, such as the Zionist movement, through political but not territorial, such as the Communist International, to highly specialized bodies such as the International Court of Justice.

### Activating the Tagil model: political conflict theory

The Tagil model provides a useful framework for the analysis of a boundary conflict. But the model still needs a trigger for a conflict to activate it. Political conflict theory provides some ideas about the origins of the triggers of boundary conflicts.

One theorist, Connor (1972), takes the view that the key trigger is always some particular contact or set of contacts. For example, economic and social changes may bring similar ethnic groups on either side of the state boundary together into regular contact, leading to the development of a new ethnic consciousness across state boundaries with demands for secession, boundary changes, and so on. This idea has been developed by Weiner into a substantial model with sixteen components, each reactive upon the other and potentially reinforcing the other, but each also having the capacity for autonomous change.

Weiner's model (1971) requires in the first instance some cross-boundary contacts along the lines of those discussed by Connor leading to the emergence of an irredentist state (which is one which seeks to create a larger state to incorporate ethnic minorities in other states with which it feels it has some affinity):

1. The irredentist state seeks to form alliances against the anti-irredentist state which is on its boundary and contains the ethnic minority with which the irredentist state feels some affinity.

2. The threatened anti-irredentist state answers by seeking to form alliances of its own.

3. Neighbouring states and the great powers are drawn into the conflict through appeals by one side or the other or through pressure from its own citizens.

4. When the irredentist state becomes interested in (for example) the ethnic minority in the other state, that minority can get the idea that it will be annexed to the irredentist state or achieve national sovereignty in its own right; this may be a quite new way of thinking for that minority – it may never have occurred to most of the minority before.

5. The minority in question may react to the irredentist demands in three different ways: it may accept the existing boundary and try to improve relations between the neighbouring states, it may support the demands of the irredentist state, or it may decide to go for independence.

6. If the irredentist state persists, the other state will tend to question the minority's loyalty more and more.

7. The irredentist state becomes sensitive to all measures, such as security measures, that affect the minority in the other state.

8. The more the irredentist state works for a revision of the boundary, the more that question draws attention away from domestic matters in the irredentist state, which may be politically convenient. High priority for the irredentist policy favours those within the country who demand domestic order, as well as unity and militancy in international relations. Appeals for 'national unity' can be used to stifle popular opposition to domestic oppression.

9. When emotions are stirred up without the irredentist demands being fulfilled, the irredentist state's elite and population are inclined to take risks in foreign affairs without considering the chances of success; they have raised the domestic as well as international political stakes. The highest stakes are played if the irredentist state arms refugees or people living near the boundary against the neighbouring state.

10. A crucial factor affecting the outcome of all the action so far is the integration of the two states. Are they nation-states as defined in Chapter Three (pp. 91–6)? Both states use intensive propaganda on their own population, on the minority in question and on the population of the opposing state. As far as their own populations, the objective is to strengthen national consciousness.

11. Opposition politicians within each group of actors come to be considered disloyal.

12. Finally, the conflict may be settled in a number of ways, all of which involve violence or compulsion, or at least the threat of it. Least likely at this stage is a resolution of the conflict by means of voluntary agreement between the parties.

(from Weiner 1971; Tagil et al. 1977: 101–2)

The Weiner model was built around the idea of boundary conflict arising from ethnic conflict, but it would work equally well with other wider reasons for conflicts between states. Boundary conflicts are, after all, usually a secondary effect of actions designed to achieve other goals. In this book, for example, Hitler's expansion of Germany has been put under the heading of 'Power through might' and the objectives of the expansion have been discussed in terms of a wider philosophy, but a series of boundary conflicts and interest groups with a direct bearing on boundary conflicts were soon formed in Germany, Schleswig-Holstein, and Czechoslovakia. The boundary is the legal mechanism which plays a central role in the enactment of broader conflicts of power; it can be central to a conflict or it can be peripheral; it nearly always influences the progress of the conflict. It is rarely the prime cause of conflict but it is frequently the prime focus.

Boundaries provide territorial claims with clarity. They help to define and stabilize the state. They can provide an obvious and tangible focus for a disagreement or dispute. The legality of a boundary is a more definite object to fight about than might, right, and nationhood. Boundaries come to symbolize the underlying causes of conflict; they are a powerful combination of symbolism and simplicity.

## The nature of state boundaries

### Introduction

State boundaries mark the limits of sovereign authority and define the spatial form of political regions. They also have a profound impact on the human landscape through which they pass. Economy, society, culture, and psyche are cut by state boundaries; trade may be cut, social exchange may be stopped, the languages taught in schools may be different, and misunderstandings between people on either side are likely. From the Oder–Neisse line between Germany and Poland to the Ruvuma–Msenje line between Mozambique and Tanzania, state boundaries have shaped the history of nations and states.

The setting of a boundary starts with the allocation of land to states, which is subject to political decisions arising from the distribution of power through might, right, and nationhood discussed in previous chapters. The treaties and conferences which have created states and their boundaries started the processes of boundary delimitation and demarcation.

### Boundary delimitation

The delimitation of a boundary is the final selection of a specific line within an allocation zone. Sometimes the allocation zone is extremely narrow. McEwen (1971: 41–54) gives several examples in his study of East African international boundaries of how meridian lines were strictly adhered to despite obvious absurdities and, worse, almost unidentifiable lines such as the Senegal–Gambia boundary involved years of detailed work on delimitation in the field under difficult conditions costing many lives and a great deal of money.

By an Anglo-French agreement of 1889, part of this boundary ran on each side of the Gambia River at a distance of 10 kilometres from the river bank. The map attached to the agreement shows the course of the Gambia to be extremely tortuous and the delineated boundary on each side of the river is portrayed as a series of intersecting arcs of circles, each having an apparent radius of 10 kilometres, the centres of which lie at the middle of the major bends of the river. It is almost inconceivable that such a clumsy boundary description could have been devised.

(McEwen 1971: 47)

Accurate delimitation and subsequent clear demarcation help to reduce conflict because they remove uncertainties which can easily become specific causes of disputes. Touval (1966) illustrated this point by comparing Africa after independence with Latin America after independence: African boundaries were clearly defined whereas the Spanish empire's hazy divisions provided causes of disputes for generations. Luard (1986: 19) confirmed that between 1918 and 1968 unclear boundaries, even in the remotest and most desolate of areas, were the single most important immediate source of conflict. Clear delimitation followed by accurate demarcation locates a boundary and contributes to its character with posts, buoys, walls, or other signs; water boundaries, whether rivers or seas, present particular problems (Prescott 1985). It is, however, the administration of a state boundary which contributes most to its character, to the boundary landscape, and to its importance in the lives of those who live near it or are affected by it. The exact location of the boundary between Norway and Sweden is clearly marked and beyond dispute, so is the Iron Curtain boundary between the two Germanies; the two are, of course, very differently administered and quite different in character.

## Boundary administration

A helpful model of state boundary administration is provided by House (1982: 7–12, Appendix) in his study of the *Frontier on the Rio Grande* between the United States and Mexico. The boundary is seen as a barrier in three senses: 'habitat', 'economy', and 'society'. House makes the important point that the barrier as perceived by Americans, Mexicans, and migrants is part of the overall reality of the barrier – it has both subjective and objective qualities.

The administration of a boundary entails decisions on the flow of labour and capital across it and the management of the local environment. In the case of the boundary between Mexico and the United States, for example, the significance of the economic barrier depends to a large extent on US government decisions about the amount of resources to be devoted to preventing the illegal migration northwards of Mexican labour. Recently, much more attention has been paid to the social question of stopping the movement of narcotics; marijuana, heroin and cocaine are all smuggled in large quantities from Mexico to the United States by land and air, and more sophisticated means of interception are needed to try to slow down the movement. The growth of the drugs trade was a trigger for a severe tightening in the administration of the boundary; light aircraft were intercepted, border patrols increased, commuters and holidaymakers were searched. Drug seizures increased, but the whole character of the boundary changed as its administration changed. It became more difficult to cross it casually. The distance friction it represented increased – in other words, the difficulty of crossing the boundary increased from the equivalent of five miles extra on a journey to fifty miles: drivers were stopped and questioned, queues built up, suspicion increased. Less regular exchange of people made casual co-operation more difficult, affecting the perception by each side of the other.

Co-operation at official level was soon dogged by just this kind of mutual suspicion, even on such relatively uncontroversial matters as environmental protection. In the 1970s the habitat barrier was breaking down as the two sides co-operated on aquifer management, the spread of infectious diseases, and air pollution, but by the mid-1980s remedies for the pollution of the Rio Grande had become bogged down in fruitless discussion on pollution standards. The boundary was still no Berlin Wall, but the atmosphere had changed (House 1982: 177–98).

The people most restricted by any boundary administration are obviously those living nearby. The effects on them of changes in the way the boundary is administered can be devastating. The boundary between Kenya and the Somali Republic, for example, has recently been closed and a fence built, turning it into a completely effective economic and social barrier. Somali herdsmen used to migrate in famine years to the Northern Frontier District of Kenya, but now that route to partial famine relief is closed. At the same time, the Northern Frontier District has suffered economically from a loss of markets in the Somali Republic (Markakis 1987: 125–31).

Another example, this time of a habitat barrier, is the boundary area between North East Poland and the Soviet Union, where a closed boundary at Białowieża marks the limit of one of Poland's major national parks. Nature, of course, recognizes no boundary and huntsmen in the Soviet Union recognize no national park. There is constant ill-feeling between the Polish park administration and the Soviet local authority, and active steps have been taken on both sides for different reasons to try to stop movements across the boundary in either direction by such rarities as wolf, bison, and beaver.

Boundaries as barriers

These examples of boundary administration and its effects on the population of the boundary zone has long been of interest to geographers modelling patterns of settlement, trade, and other transactions between theoretical settlements on a plain (Haggett 1965: 59–60). The boundary represents additional distance friction in these models. The distance friction between Białowieża and the Soviet Union is several hundred miles for human beings, although the physical distance for a wolf, bison, or beaver is less than a mile. The distance friction between the West Berlin enclave of Lichtenrade and the rest of the western world is another case where a boundary artificially creates distance friction and all the social, psychological, and economic problems of remoteness. Boundary areas of this sort have the same problems as physically remote locations for industrial investment and they suffer in theory and practice a severe locational disadvantage which they need subsidies and special government help to overcome.

Boundaries are more generally significant economically because of the tariffs and trade barriers they represent. This aspect of boundary administration converts an otherwise open system into a series of artificial economic environments. In these circumstances, people can be gravely affected when boundaries are shifted. For example, Tagil's (1970) study of areas of Schleswig-Holstein (transferred from Germany to Denmark after the First World War) shows that the change inflicted considerable economic suffering for dubious political gains: under German control, the farmers in the affected areas had exported their products profitably to German industrial areas at protected rates without foreign competition; subsequently they had to compete as part of Denmark's agricultural economy based on free trade in dairy and pig products. The reverse may also occur. One example would be the farmers not taken into Denmark but left behind in North East Germany who, without competition from Schleswig-Holstein, were able to do rather well. Other examples are towns like Helmstedt (on the West German frontier with East Germany) and Frankfurt-on-the-Oder (on the East German frontier with Poland), which are towns at crossing points on boundaries with few crossing points.

Many boundary positions are selected, fought for, and enforced primarily for defensive reasons. From the historical struggle by France for a boundary on the Rhine through to the Israeli incorporation of the Golan Heights, the administration of the boundary zone

129

is frequently orientated around military garrisons. The settlements near the boundary between Israel and Syria were *Kibbutzim* (Israeli farming and industrial communes) either run by the Israeli army or with a strong military element. In this way they symbolized Israel's permanent presence and physically reinforced the boundary on the Israeli side. At the same time, military considerations have forced the closure of the Golan Heights boundary even to local Druze families, now permanently split. Military considerations often account for the closure of a boundary with all the social, economic, and environmental implications that may have.

The boundary between Northern Ireland and the Republic of Ireland is an example of military considerations directly conflicting with the economic interests of the local population. A maze of country lanes crisscross the boundary near Newry, and they are used regularly by Northern Irish Protestants and Roman Catholics alike to get to the shops in the south and by the Irish in the south to get to the north for petrol, drink, and other goods which are cheaper in the United Kingdom. The British Army, to prevent the inflow of terrorist arms and personnel along country lanes, have blocked or blown up a number of crossing points, making long detours necessary, but only until locals have dislodged or filled in whatever obstacles had been put in the way.

The nature of boundaries varies widely, depending on their delimitation and demarcation but above all on their administration. Whether closed or open, easy to cross or hard to cross, they have a profound impact on the people and places around them. These are the areas most directly affected by the legal limits and definitions of power which the state boundary represents. But, of course, by enclosing sovereign states, international boundaries affect us all.

*Case-study 4.1: the Sino–Soviet boundary dispute*

History

Few boundary lines have become such a clear focus of ideological and strategic disputes between major powers as the boundary between the Soviet Union and China (Ginsburgs and Pinkele 1978; Robinson 1970; Tsien-hua Tsui 1983). The expansion eastwards of Tsarist Russia presaged an inevitable clash with China. A boundary line would some day have to be drawn (Figure 4.2).

Up to the end of the sixteenth century, Chinese authority or suzerainty extended over the whole area that now contains the Sino–Soviet boundary, from the Karakoram to the Sea of Japan. It was towards the end of the fifteenth century that Muscovy was unified;

Ivan IV proclaimed himself 'Tsar of All the Russias' in 1547, but his empire was purely European.

By 1643, Russia had reached the Pacific. At first, the territory annexed in this expansion eastwards was just a belt of semi-permafrost in the Siberian north. Not only was this sort of country of little interest to China, it was also useless to the Russians because it produced practically no food. Russian settlers' attention naturally turned towards the warmer land to the south. In so doing, they encroached inevitably on areas of Chinese suzerainty or loose control, and these incursions were the first signs of the causes of conflicts to come.

Following a number of minor disputes caused mostly by the disorganized movement south of bands of Russian settlers, emissaries were exchanged between the Tsar in Moscow and the Emperor in Beijing, and a treaty was signed. The Treaty of Nerchinsk (1689) was the first treaty China ever signed with a western power (the wording of the treaty was in Latin with Chinese and Russian translations). It established, with only a few subsequent revisions favourable to the Russians, the modern boundary from the Altai Mountains to the sea, between Russia and China and between Russia and Mongolia. The rest of the boundary was agreed by a series of treaties between the Treaty of Kyakhta in 1727 and the Treaty of Tarbagatai in 1864.

After the October Revolution of 1917, the new Soviet government declared that it would renounce conquests made by its Tsarist predecessor in China and agreed to give up Russian interests in Manchuria. It was to be the one and only concession, for by 1921 the remnants of the White Russian Army resisting the Soviet Red Army were making their last stand under Baron Niklaus von Ungern-Sternberg in the Outer Mongolian Republic which had declared unilateral independence from the Chinese Empire in 1911. The Red Army moved into Mongolia in 1921, defeated the White Russians, and helped the local Communist leader, Sukhe Bator, to found the People's Republic of Mongolia (the world's first people's republic) in 1924. This was to be an independent state, but it was always run along Soviet lines with a Soviet military and political presence. The virtual incorporation of this nominally independent state into the Soviet Union was effectively a major shift of the boundary south from the lines agreed at the Treaty of Nerchinsk.

Chinese support for forces opposed to the October Revolution meant that any further negotiations after the withdrawal from Manchuria were put off and the Soviet military presence in Mongolia was not to be negotiated. Various Soviet moves reinforced the new boundary; for example, troops were moved to the Mongolian border

131

at Dzamin Uud, more than halfway from the Soviet Union to Beijing. Additionally, the Uranghui region of Mongolia, to the west of the People's Republic, was reinforced and also given nominal independence while occupied by the Red Army; indeed, unlike the People's Republic, it had no independent forces of its own – the Uranghui Republic of Tannu Tuva was actually incorporated into the Soviet Union in 1944, whereas the People's Republic of Mongolia has remained independent. There were also various Soviet incursions across the Altai boundary into the far west of China but the modern boundary line, agreed at Tarbagatai in 1864, was agreed again when the Communists took power in China in 1949.

Relations between the Soviet Union and the People's Republic of China were initially very good, and ten years of co-operation from 1949 were without boundary incidents. In 1959, however, there was a major revolt against the imposition of Chinese Communism in Sinkiang and there appeared to be considerable Soviet involvement in fermenting agitation. Some 50,000 refugees crossed the boundary from China into the Soviet Union in the next three years, and it was the start of an acrimonious period in Sino–Soviet relations to which the boundary dispute contributed but which had wider causes. At its worst, there were bloody skirmishes on the boundary, such as when thirty Soviet troops and an unknown number of Chinese were killed fighting for Damanskiy (or Chenpao) Island in the Ussuri River, 100 miles upstream from Khabarovsk, on 3 March 1969. On the Amur River, the Soviets claimed complete territorial control, including both banks, on the basis of rather vague maps, associated with a revision of the Treaty of Nerchinsk at St Petersburg in 1881; and in July 1969 there was the Battle of Goldinskiy (or Pacha) Island in the Amur River. The Soviets started to concentrate troops in the boundary area and the Chinese started a massive nuclear shelter building programme across the country.

War never came. Neither China, with vastly inferior forces, nor the Soviet Union, overstretched militarily and economically, ever found it in their interests to embark on anything worse than border skirmishes.

The whole dispute over the boundary reflects and symbolizes the clash between Stalinist and Maoist views on the importance of territorial boundaries and the role of territory in defence. It also represents the clash of different political philosophies and the geopolitical strategies arising from them. These two clashes are considered in turn, as both were expressed physically on the Sino–Soviet boundary.

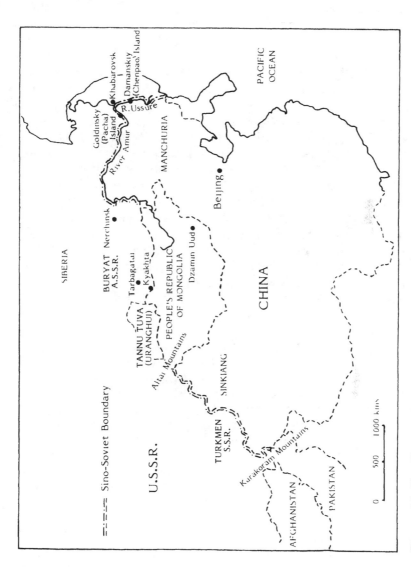

*Figure 4.2* The Sino–Soviet boundary

### Opposing views on the importance of territorial boundaries

Stalin took the view that territory was the single most crucial factor in military defence right from the days when there were Far Eastern and Northern threats to the October Revolution itself. As Commissar of Nationalities and General Secretary of the Soviet Communist Party, he made sure that none of the constituent Soviet Socialist Republics of the Union of Soviet Socialist Republics would take advantage of their constitutional right to secede from the Union. Republics had this right so long as they had some share in the Union's international boundary. Stalin made sure that any doubtful cases, where the Republic's elected Soviets were unreliable, would either remain technically independent but be occupied by the Red Army (like Tannu Tuva) or remain Autonomous Republics without the right to secede, such as the Buryat Autonomous Soviet Socialist Republic. Another alternative was to ensure heavy settlement by ethnic Russians to counterbalance possible Moslem solidarity, as in the Turkmen Soviet Socialist Republic. All these methods were used around the disputed boundary with China. They were buffer territories where, albeit by organizational guile, the local population or the representative had expressed a wish – or could be said to have expressed a wish – to stay in the Soviet Union (Deutscher 1966; Whiting 1957).

The view of the Chinese Communist leader, Mao Zedong, clashed with Stalin's. As far as defence was concerned Mao's model was to fight an invasion by drawing the enemy in deep: large tracts of supposedly secure territory were not significant. Territory could symbolize grandeur and success and it could contain important resources, but it had limited military significance. Mao's view of the Chinese claim to territory was therefore not based on a perception of defensive needs, which had been Stalin's main consideration, and Mao was not concerned with local democratic preferences; he was instead preoccupied with historical claims. He rejected popular sovereignty and self-determination in favour of territorial statehood expressed in anti-colonial legal and constitutional terms, which was typical of the post-colonial Third World (Medvedev 1986: 5–65; Terrill 1980).

Mao Zedong claimed that the boundary treaties which had been signed with the Russians and later with the Soviets lacked legitimacy because they were drawn up at a time of colonial exploitation. The argument was that the treaties between the Chinese Empire and Tsarist Russia had been unequal and therefore should not be recognized. Mao took the view that international law provided that treaties had to be concluded on a reciprocal basis in order to be valid; they had to apply to states of sovereign equality, which, he claimed, was certainly not the case with any of the treaties with the west, including Russia, signed by a crumbling Chinese Empire from the seventeenth century onwards.

Mao's argument was met with a blunt rejection of the whole concept of the unequal treaty by the Soviet Union: 'References by Chinese propaganda and Chinese diplomacy to the allegedly unequal character of the old treaties [are] politically inadmissible even in the event they are historically accurate' (Khvostov 1964 quoted in Tsien-hua Tsui 1983: 94).

The Sino–Soviet boundary dispute can then be seen in part as an expression of two conflicting concepts of territory in two contrasting political models. On the one hand, there was Stalin's model based on the centrality of territory in defence and his consequent cynical use of self-determination. On the other hand, there was Mao's rejection of the defence argument and his adoption of legal and historical argument as a means of regaining territory as a national resource and for national pride. More cynically, Mao's emphasis on the idea of the unequal treaty made it possible to portray China as continuing an unequal struggle gallantly against an imperialist aggressor; there is, however, no particular evidence that the Chinese have used the boundary dispute with the Soviet Union as a diversionary tactic during domestic crises.

Opposing values

The Sino–Soviet boundary dispute can be seen as an inevitable consequence of the history of the boundary itself, plus the clash of Stalinist and Maoist political models with their two different and incompatible objectives. Following Tagil et al. (1977: 121–30), it may now be possible to go beyond the direct historical reasons and beyond the indirect conceptual reasons for the boundary problem and to find a further cause of the clash in the dominant values of the two sides. In the case of the Soviet Union, the dominant values were cultural and political ones associated with Eurocentric Marxism-Leninism. In the case of China, the dominant values were, for ten years after the Communist revolution, similar ideological values to those of the Soviet Union but associated with Mao Zedong's interpretation of Marxism-Leninism; later these changed to nationalistic values associated with the geopolitical fear of encirclement.

Lenin was for the most part Eurocentric. His and Stalin's view of the anti-colonial struggle concerned mainly the contribution which that struggle could make to the crisis of capitalism in developed states which, in turn, would eventually bring about the downfall of capitalism across the world. For example, Lenin welcomed the 1912 Chinese revolution, as earlier Marx had welcomed similar less

135

severe uprisings, because it could divert resources and cut supplies and markets, hastening the victory of the proletariat in developed states.

In Lenin's view, excess profits from exploitation of the colonies allowed the metropolitan bourgeoisie to stave off revolution at home by buying off the upper strata of the proletariat, giving them a share of surplus wealth and thereby splitting them off from the rest of the proletariat. The elimination of some of this surplus as a result of colonial revolution would end this process of bribery and division, opening the door to socialist revolution in the metropolis. This of course could be extended from traditional colonialism to modern neo-colonialism, economic exploitation of former colonies without direct political rule (Lenin 1948).

Despite his emphasis on the importance of national liberation to the proletarian revolution, Lenin never conceded equality or primacy to national liberation. The world revolutionary process could only be carried on by the proletariat in capitalist states. He believed the Russian working class and the Communist Party of the Soviet Union to be in the vanguard of the international proletariat. The Chinese peasant and the Chinese Communist Party had only a diverting role in economic thought. Theirs could not possibly have been the proletarian revolution they kept claiming it to be.

The Chinese Communist revolution saw itself as a Marxist class revolutionary struggle and not as an anti-colonial struggle, and certainly not as a subsidiary to class struggle elsewhere. As such, it was committed to the principle that, through socialist internationalism, only socialist states could effectively help each other to create a demonstrably economically powerful bloc in competition with capitalism. For ten years after the Chinese Communist revolution of 1949, relations between China and the Soviet Union were indeed internationalist. Development aid poured into China in quantities which were quite astounding given the economic conditions of the Soviet Union at the time. It was inevitable, however, that the relationship would break up as ideological difference emerged.

Towards the end of the 1950s, doctrinal changes in Soviet Communism hastened the split. The increasing willingness of the Soviet Union to compromise with the capitalist world was combined with an increasing reluctance on the part of the Soviet Union to supply top-quality weaponry to China, including an absolute embargo on nuclear weaponry. This rejection of socialist internationalism offended China's national and revolutionary pride. The failure of internationalism forced Mao Zedong to adopt policies suited to China alone with only minimal regard to international revolution. The new policies were to include a new national cultural revolution and a revival of the old Chinese national political fear of encirclement by hostile powers. From inspiring ideology, Chinese foreign policy degenerated to dealing with the predictable concerns of the state.

Mao's successors have been preoccupied in their foreign policy with the Chinese state and signs of its encirclement. They see a Soviet threat in Vietnam, Kampuchea, Afghanistan, and Mongolia, as well as along the Sino–Soviet boundary itself (Yahuda 1983: 188–233). The Soviet government under President Gorbachev is taking steps to allay China's geopolitical fear by organizing withdrawals in both Afghanistan and Kampuchea. At the same time, in a major policy speech in 1986, Gorbachev specifically accepted that the boundary should run along the main channel of the Amur and Ussuri rivers which would even involve surrender of islands bitterly fought over in the past. In addition, he has already withdrawn some 15,000 of 70,000 troops from the People's Republic of Mongolia.

The Chinese no longer seek ideological objectives and their geopolitical and legal–territorial claims may now be met by the Soviet Union. The values behind Soviet policy are also changing. Gorbachev is primarily concerned to save socialism in the Soviet Union from economic disaster: the value of territory for defence, stressed by Stalin, had been redefined to take account of the extent to which it is economically and militarily controllable. The clashes of values and of political models are disappearing. If the wider causes of the dispute go, no focus is needed; so it is no surprise that a solution to the Sino–Soviet boundary conflict is at hand.

*Case-study 4.2: a federal boundary (Britain–Channel Islands)*

The status of the Britain–Channel Islands boundary

Boundaries within federations of states can be like international boundaries in some respects or like mere parish boundaries in others. Studies by Duchacek (1986) and Verney (1986), for instance, illustrate the wide variety of federal boundaries. So long as neither state or sub-state on either side of the boundary is trying to assert itself militarily or in any other way over the other and so long as there are substantial shared interests, it can be a boundary of convenience, strengthening or weakening, closing or opening, whenever or wherever it suits both sides. An example is the boundary between Britain and the Channel Islands, a vital but imaginary line across the English Channel.

The Channel Islands have long held a unique constitutional position in relation to Great Britain. They are possessions of the Crown, and the British government is responsible for their defence and the

137

conduct of their international relations, but they are not part of Great Britain or the United Kingdom. The constitution and some of the laws of the Channel Islands are based on the ancient customs of the Duchy of Normandy, of which they formed a part until 1204, when the continental part of Normandy was lost to France. The Queen is sovereign only because she is the successor of the Duke of Normandy and not because she is Queen of England. The Channel Islands are neither colonies nor independent states; they enjoy a wide degree of autonomy in domestic policy, including financial autonomy. They are subject to the legal supremacy of the Westminster parliament in which they are not represented, but by a long-standing convention it is required for Westminster to legislate separately for the islands and, where this is intended, it is usual for parliament to ensure the support of locally elected representatives (Royal Commission on the Constitution 1975, 6: 371–7).

There is then a federal boundary between Great Britain and the Channel Islands. There are four main factors which strengthen it: distance, political dissonance, mutual economic advantage, and cultural insularity; and there are four main factors which weaken it: political concurrence, the conduct of international affairs, migration, and unviability. (The specialized information in this section is from a series of unpublished interviews with officials and residents of the Channel Islands and responses to questionnaires which were part of a research project supervised by the author and carried out by L. C. Gainham.)

Factors which strengthen the boundary

The most obvious factors which strengthen the boundary between any two places are the boundary's characteristics as a physical barrier. One hundred miles of sea between Great Britain and the Channel Islands has its effects. For example, it is inevitable that telecommunications and electricity are independently run in the Channel Islands; it is the only economically sensible way to run them. In the case of electricity there are no links with the British National Grid, although there is in progress a feasibility study concerning the possibility of establishing a link with France.

The main sources of political dissonance between Britain and the Channel Islands are health care and social services. Most health care and social services are free to British citizens and are funded by central government, but the Channel Islands have been persistently opposed to any such provision. Provident societies and friendly associations operate a number of hospital and medical insurance schemes on the islands and residents are advised to take out insurance. Politically defiant, the Channel Islands have rejected the

British scheme time and again, and health care provision on the islands is much more in line with the system in the United States than it is with the British one. As the social structure is different from that of Britain, private health care is seen to be a much more efficient way of managing resources – the majority of islanders can afford insurance with reasonably good cover.

A clear boundary between Britain and the Channel Islands is to the economic advantage of both sides. The Channel Islands levy their own taxation; income tax is at a very low rate and there is no inheritance tax or any other form of death duty. Consequently, the Channel Islands derive a considerable benefit from their activities as a financial centre. Being in monetary union with Britain, their substantial foreign currency earnings are of value to both the islands' and the British economies. There is, for example, significant revenue from companies registered on the islands, and this makes a contribution to the British balance of payments.

British companies have also found it extremely useful to have offshore companies through which they can conduct delicate business, for example, in involving arms for South Africa and high technology goods for the Soviet bloc. There has even been talk of the islands housing some of the 'Marcos Millions'!

The cultural insularity of the Channel Islands is exemplified by their housing rules. To be a resident on one of the islands there is no nationality qualification for British citizens, but there are strict housing rules. Under the terms of one of the local housing laws on Guernsey,

> the law provides that a person who is not a qualified resident
> may not occupy a dwelling other than a registered dwelling with
> a licence from the housing authority. . . . If a licence holder
> wants to move to a post with another employer, he should make
> application to the authority while his licence is still valid and
> before resigning or committing himself to another post.
>
> (States Housing Authority 1982)

A person from Britain will only be employed on the islands if there is no one there to fill a particular vacancy, but even then a housing licence is required if the employee is intending to live there. This means that, once a term of employment has been completed, the licence can be revoked and the former employee told to leave. Only those born on the islands or who can afford to buy their own home – property prices are very high – are really welcome. The islands are exempt from the EEC rules on freedom of movement.

Factors which weaken the boundary

Turning to factors which weaken the federal boundary, the first of these is political concurrence between the two sides. The Channel Islands and Great Britain have both expressed the same aims on such matters as immigration and defence, and the Channel Islands are entirely in the hands of the British government on both of these issues. As far as immigration is concerned, the islands are in the British Common Travel Area, so local immigration controls have to be compatible with those which are exercised at mainland ports by the British Immigration Service. Where the Channel Islands do have their own laws, they are identical to British laws and only reinforce enthusiastically Westminster restrictions (as one would expect from their restrictive housing rules). As far as defence is concerned, the islands make no contributions and Westminster takes all responsibility. In practice, defence policy is accepted in the islands and is, like immigration policy, enthusiastically endorsed. For example, following the Falklands War of 1982, the islands' Civil Defence Committees made a voluntary donation of £250,000 to the British government.

The conduct of international affairs from Westminster also weakens the boundary, especially as economic affairs are increasingly regulated by supranational authority. When Britain joined the EEC, for example, there had to be special clauses to safeguard the position of the islands so they would not be subject to the rules covering the free movement of people, capital movements, and the harmonization of taxation and social policies; but all other EEC laws were still applied. The Departments of Agriculture and Fisheries on the islands are obliged to observe quality requirements set out in EEC law and to abide by European rules governing competition; some of Jersey's dairying and Guernsey's glasshouse industry have been especially hard hit by competition from Europe.

The migration of British citizens to the Channel Islands has had an opposite and counterbalancing effect to local cultural insularity. The Norman French traditions of the islands scarcely exist except in artificially maintained folklore or in official documents. Emigration by native islanders for education, work, and marriage are balanced by immigration by English people for retirement or to save on taxes. The emigration and the immigration have combined to dilute severely the islands' cultural identity and separateness from Great Britain.

For some aspects of independence, the Channel Islands are just too small and unviable; in these cases, a boundary drawn around the islands is simply unrealistic. There are some facilities which the islands are too small to provide, including training for many jobs.

The islands' electricity boards, for instance, have their personnel trained in Britain where much more advanced facilities are available and training courses can be run economically for large numbers of people. Electricity administrative and technical staff are trained at British Area Boards and the British Electricity Council. The same is also true of telecommunications; each year seventy or eighty technical staff attend various training courses in Britain at the British Telecom colleges and at equipment suppliers' premises. Police officers from the island forces also receive specialized training with the Metropolitan Police in London. Indeed, British police forces give assistance – and sometimes virtually take over – on such matters as corporate fraud, drug dealing, and terrorism. At the same time, police forces on the Channel Islands have access to central finger-print libraries and the Metropolitan Police central computer. The islands are too small to have their own higher education facilities, specialized medical facilities, and many other goods and services for which a population of 135,000 is inadequate.

## Conclusion

An imaginary boundary in the English Channel between the Channel Islands and the United Kingdom may not seem at first sight the most obvious example of a federal boundary, but it does illustrate well the factors which strengthen and weaken a boundary. On the one hand, there are the factors which increase the political distance represented by the boundary and, on the other hand, factors which bring the two sides closer together. The physical distance obviously remains the same, but the political, economic, cultural, and social distance can be stretched or shrunk by policies and attitudes. The great advantage of a federal boundary between two places which co-operate together, is that it can be used as an instrument of policy, to be strengthened or weakened depending on whichever approach seems to produce the most beneficial results at a particular time.

*Case-study 4.3: Western Sahara*

## History

The most immediately striking aspect of the Western Sahara, south of Morocco on the West Coast of Africa (Figure 4.3) is that it is very hot, very dry, subject to terrible sandstorms, and with an almost harbourless seaboard of cliffs. A mainly nomadic people lived there in pre-colonial times, occupied with trans-Saharan trade by camel. They formed a distinct ethnic group of mixed African, Arab, and Berber ancestry, speaking the Hassanya dialect of Arabic.

141

Although no territorial state existed and there were no fixed boundaries, Saharawi people were distinct as long ago as the tenth century when it was a Saharawi, Yahya ibn Ibrahim el Gadeli, who founded the short-lived Almaharid Empire which stretched from modern-day Ghana to Spain. There is no evidence that the Saharawi were ever subservient for any length of time to the Sultans of Morocco or to the Emirs of Mauritania. Nominal Spanish rule in the colonial period made little difference to Saharawi life until its last twenty years. A British government report of 1919 concluded:

> A lack of fresh water and the consequent infertility of the soil, lack of harbours on the coast, the instability of the native population and of trade, and the laxity of government control, combine to make capitalists shy of investing money in this tract of country.
>
> (Hodges 1983: 51)

One thing that was quickly achieved by the Spanish colonists was a clear set of boundaries for their new Spanish territory, agreed by a series of treaties with the French who controlled all the surrounding territory, between 1886 and 1912, although some parts of this defined territory resisted occupation by Spanish troops right up to 1934 (Hodges 1983: 27).

In the 1960s substantial deposits of minerals, especially phosphates, were found and their exploitation led to major social and political change. There was large-scale investment and the phosphate industry boomed; basic infrastructure improved and small settlements grew into towns. The availability of jobs, trading opportunities, and educational and medical facilities, along with a period of drought between 1968 and 1974, all played a part in encouraging native Saharawis to abandon their traditional nomadic life-styles. By 1974, the urban population had reached over 40,000 (55 per cent of the population), trebling since 1965. It was this coming together of the Saharawis into towns which facilitated the development of the idea of a Sahara state, the political aspiration normally called 'nationalism', not to be confused with the concept of 'nationhood' discussed in Chapter Three. A state on the exact territory of the Spanish colony fitted the declaration by the Organization of African Unity that former colonial units should be the main determinants of new state boundaries (Shaw 1986: 123).

Greater Morocco (and Greater Mauritania)

By the time the Saharawi people were becoming aware of themselves and their potential to form a state, both Morocco and Mauritania were claiming Western Sahara, then Spanish Sahara. The Moroccan

142

claim rested on the idea of a historic Greater Morocco. Maps were published depicting Morocco incorporating a vast portion of Algerian Sahara, the whole of Western Sahara, Mauritania, and a corner of Mali. At first all this was met with surprise, even in Morocco, but the idea soon caught hold. The vision of a vast and glorious pre-colonial Morocco quickly captured the hearts of the newly independent Moroccan nationalists. In 1958, King Mohammed V declared that Morocco would do everything possible to recover Spanish Sahara and he ordered the claim to be lodged at the United Nations at the end of 1958.

Morocco based her claim to the Western Sahara on historical links with the territory. She talked in terms of a permanent and peaceful Moroccan presence extending back for centuries, including the appointment of sheikhs, the collection of Koranic taxes, and the military and commercial protection of the area and its people. This notion was at best dubious. Only one definite historical instance of an ancient connection has ever been demonstrated; that was when Sultan Saadians sent soldiers to impose his authorities in the oases in the area in 1581; most died of thirst and hunger in the desert and nothing was gained (Hodges 1983: 27). No doubt there were other occasions, and it was certainly the case that European states did negotiate with Moroccan authorities over the territory. When they did so, however, the Moroccans frankly admitted their lack of control, as they did when the Spanish were granted fishing rights in 1776:

[The people of the Western Sahara] are greatly separated from my dominions and I do not have any authority over them . . . . They have no fixed abode and move around as they please without submitting to government or any authority.

(Quoted in Hodges 1983: 31)

Greater Morocco seems really to have been created in the late 1950s in a mood of post-independence euphoria.

At about the same time as the Moroccans called for a Greater Morocco, the Mauritanians to the south called for a Greater Mauritania which was also to include the Western Sahara. The Mauritanian claim was based on cultural and ethnic links and only really sought the territory to combat a perceived territorial threat from Morocco on Mauritania herself. It was never a convincing or viable claim.

## The struggle for Saharawi independence

It is probable that if the Spanish Sahara had gained independence in the 1950s, the Saharawis would have become Moroccan just as the

143

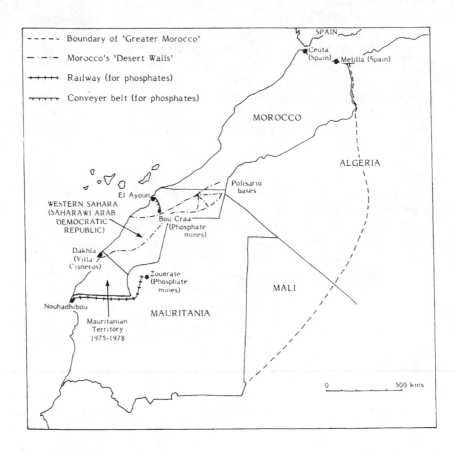

Boundary of 'Greater Morocco'

Morocco's 'Desert Walls'

Railway (for phosphates)

Conveyer belt (for phosphates)

SPAIN

Ceuta (Spain)

Melilla (Spain)

MOROCCO

ALGERIA

Polisario bases

El Ayoun

WESTERN SAHARA (SAHARAWI ARAB DEMOCRATIC REPUBLIC)

Bou Craa (Phosphate mines)

Dakhla (Villa Cisneros)

Zouerate (Phosphate mines)

MALI

Nouhadhibou

MAURITANIA

Mauritanian Territory 1975-1978

0        500 kms

*Figure 4.3* Western Sahara

Tuaregs became Algerian. During the next twenty years, the development of the towns and the development of Saharawi nationalism
led to demands for independence. By June 1970, the Saharawi had
begun to show open defiance of Spanish rule with violent anti-
colonial demonstrations which were violently put down. A militant
guerrilla group, the Polisario, was set up with a policy of armed
insurrection, and it soon carried out a number of successful hit-and-
run armed raids.

The Polisario also started to plan radical policies for independent
government. All caste, status, and tribal loyalties would end. Slavery
would be abolished. Economic resources from phosphates would be

144

distributed fairly by the state. A social plan would be inaugurated to provide adequate housing, health facilities, and schools. Education would be compulsory and would express Koranic virtues but with a strong commitment to the emancipation of women. These policies made the Polisario popular among the young and they were soon front-runners for the formation of a post-independence government. Both the United Nations and the Organization of African Unity soon agreed that the indigenous Saharawi had the right to determine their own political future by a referendum which would ask them how and by whom they wished to be ruled. Spain agreed to a referendum but would not include the Polisario. The referendum was delayed, then cancelled. The United Nations meanwhile passed a resolution insisting on independence and self-determination (Franck 1987: 13).

At that point, in October 1975, Morocco invaded. The scene was set for a bloody conflict between nationalists aspiring to a state behind the old colonial boundaries and imperialists who sought to establish hazy replicas of pre-colonial empires. It was to be a war between two sets of people claiming a right to own the same piece of land, one through self-determination within legally defined colonial boundaries and the other through armed might to assert control over a much vaguer area. In the end, as the following brief history will show, the clear legal boundaries prevailed.

Morocco in the early 1970s was going through a time of political turmoil and economic difficulty. Two recently attempted coups and popular riots led King Hassan to seek some major distraction. A crusade to recover part of Greater Morocco would help restore royal prestige; people would stop rioting and the army would find new loyalty in this grand scheme.

Although Spain had 20,000 troops in the territory and the decision of the International Court was still awaited, King Hassan's timing had been perfect. The United Nations had no clear view and no wish to get involved. Spain, at the tail end of the Franco regime, had no intention of becoming involved in a war in Africa, and also feared revenge on her African enclaves of Ceuta and Melilla and on the 18,000 Spaniards living in Morocco and Spanish business interests there. The United States would not oppose a strategically located pro-western regime. As Hassan's 'Green March' progressed, Spain withdrew from her colony without firing a short.

On 14 November 1975, Spain, Mauritania, and Morocco signed the Madrid Accords which granted Spain fishing rights off the Saharan coast and certain mineral concessions. Morocco and Mauritania were to share the former colony itself.

Spain finally withdrew in February 1976 and Morocco went to great pains to win the loyalty of the Saharawis. To improve economic conditions, tax-free zones were created to encourage Moroccan business investment and a huge development programme was launched. By 1985 it was estimated that practically a billion dollars had been spent. Morocco even signed a ten-year contract with Club Mediterranée to help develop tourism. But the situation in the territory soon became chaotic due to the effective guerrilla campaign waged by the Polisario. Most Saharawi live in exile in refugee camps in Algeria to which they fled at the time of the Moroccan invasion. Only some 40 per cent stayed behind, and the main beneficiaries of Moroccan investment turned out to be Moroccan immigrants.

The Saharan Arab Democratic Republic was declared in exile in February 1976 and was soon recognized by seventy states, including most members of the Organization of African Unity which bases its view of legality on colonial boundaries at the time of independence. For the Saharawi who remained in the Moroccan colony, life soon became harsh and Amnesty International reported a number of 'disappearances' (Amnesty International 1985). At the same time, prosperity declined as the Polisario damaged the infrastructure associated with the phosphate mines.

Guerrilla tactics made the Polisario an elusive enemy. At first they concentrated on Mauritania and forced her out of the war. Using long-range raids, the Polisario could easily penetrate deep into Mauritanian territory. These raids were unstoppable by Mauritania's small army, and soon most of Mauritania's own phosphate industry – nearly all on the Western Saharan boundary – was at the mercy of the Polisario. Mauritanians soon became disillusioned by the economic hardship, and they never had the same nationalistic fervour for Greater Mauritania as Moroccans had for Greater Morocco. Eventually Mauritania employed 8,000 Moroccan soldiers and even had direct help from the French Air Force against Polisario bases, but even the French could not stave off the inevitable military coup of July 1978. Mauritanians wanted peace and stability and the new regime ordered a withdrawal from Western Sahara and dropped its claim to territory (Kamil 1986: 36–51).

For Morocco, the war against the Polisario turned into one of prolonged attrition, causing social, economic, and political tension within Morocco itself. The huge military expense cost the Moroccan economy dear. The sheer increase in the size of the army imposed a heavy burden, and debt accumulated as more and more expensive imported military hardware was needed. To stop Polisario raids, an expensive desert wall-building has been undertaken since 1980 in an attempt to create safety zones within Western Sahara while retaining only nominal control of the rest of the territory. At one time, over one-third of Moroccan GNP was spent on arms and on building desert walls.

There has been civil unrest in Morocco as a result of the economic burden of the war emerging in the form of poverty, inflation, and unemployment. There has also been discontent in the Moroccan army. After 1978 and the withdrawal of Mauritania from the war, the Polisario was able to concentrate on Morocco. Before the building of the walls, Moroccan troops were subject to constant demoralizing ambushes, but still in 1988 Moroccan soldiers defending the desert walls felt like sitting ducks waiting to be killed at the will of the Polisario. Conditions were bad and there were some local mutinies (Turner 1986: 20).

Algeria has been the most ardent foreign supporter of the Polisario. The Algerian leftist government had its own separate territorial dispute with Morocco as well as a suspicion of the right-wing Moroccan royalist regime. At the outset, Algeria gave the Polisario her support in the form of 'total and unconditional commitment to the National Liberation struggle of the Saharawi people which they are waging under the leadership of the sole legitimate representatives of the Polisario front' (Hodges 1983: 331).

Algeria's role has been vital to the Polisario by putting political pressure on Morocco and by providing a Saharawi place of exile, as well as arms, military training, and bases for the Polisario. The consequent tension between Morocco and Algeria caused world concern, never more so than when Morocco signed a military agreement with Algeria's other neighbour, Libya.

By causing such substantial international problems as well as disrupting the economy of their two neighbours who claim their territory, the Polisario have finally won the Saharawi right to self-determination. On 30 August 1988, a United Nations' proposal for a referendum was agreed by both sides. The United Nations also agreed to put a military governor in charge of the referendum and in charge of the whole territory until the referendum in spring 1989. The right to vote is to be restricted to the people who were in the last Spanish census in 1974, which will avoid Polisario suspicion that Moroccans who moved in during the Moroccan occupation would vote, and it will avoid the Moroccan suspicion that 8,000 Polisario would flood in with assorted malcontents in its wake.

The choice in the referendum is to be between total independence, self-rule under Moroccan sovereignty, and continuing as part of Morocco. Agreements reached at a regional summit in May 1988 between Algeria, Libya, Mauritania, and Morocco will encourage the second of the three choices. King Hassan now says he would be satisfied with mere titular sovereignty, and some compromise which would combine such sovereignty with the return of a Polisario government may well ensure peace for some years (Soudan and Sada 1988).

On the other hand, it is hard to imagine the Polisario giving up their aim of sovereign statehood within the old colonial boundaries after a decade of ultimately successful struggle. Considering how few people there are in Western Sahara and how great are its mineral resources: it may well be that the Saharan Arab Democratic Republic will be one of the more truly independent small states with the potential to be a nation-state, exercising real self-determination in its affairs and real control of its land.

## Case-study 4.4: *the end of the partition of Jerusalem*

### History

> Ten measures of beauty came down: nine were taken by
> Jerusalem and one by the rest of the world.
>> (Talmud: Kiddushim, 49 BC, quoted in Grindea 1982: 175)

Jerusalem was the last part of the Promised Land to fall to the ancient Hebrews, the last and most prized possession in the Land Flowing with Milk and Honey. King David captured it from the heathen Jebusites and, from Jerusalem, he united the kingdoms in the north and south, and established a great empire from Damascus to the Red Sea.

Some of the more ardent Zionists of modern Israel take the view that David's empire should be revived (to include, incidentally, Philistine territory on the Mediterranean coast rather than Damascus). In 1947 the United Nations voted for Jerusalem to be an international zone, shared between Jews and Arabs with neither side having sovereignty; but by the end of the Israeli War of Independence in 1949 the city was partitioned between the two sovereign states in occupation, Israel and Jordan; the local Arabs had no say, not even over the Arab quarter which came under Jordan.

In 1967, Israel conquered what had become Jordanian Jerusalem, as well as the West Bank, the Golan Heights, Gaza, and Sinai, from the Arabs. Everything was ultimately negotiable for peace except Jerusalem. For practically all Israeli Jews, the city was irreversibly reunited, irrevocably owned and controlled by the Jewish state, forever its capital. Unlike any other territory captured, Jerusalem was incorporated immediately by the Israeli parliament (the *Knesset*) by a constitutional amendment into the Israeli state. What remained to be done was to remove the boundary in practice and not just in theory. The physical boundary, the wall, locked gates, and gun emplacements, went at once in 1967, but the partition remained in all but name.

Attempts at functional unification

The immediate administrative decision by Israel, the new master of all Jerusalem, was to try to work with any willing Arab authority. Unelected village elders or *Mukhtars* from each of the main Arab suburbs of Jerusalem were the most co-operative at the time and, in most cases, formed a fairly close relationship with the Israeli municipal officials who were put in charge of reuniting the city in practical matters at local level. These important officials were given the following instructions:

> In matters which relate solely to Arabs, act as though you were an Arab. In any Israeli–Arab confrontation, try to mediate and compromise. Be a loyal representative of the Israeli authority and its policies among the Arabs, but at the same time be a loyal representative of the Arab view before the Israeli authorities.
>
> (Instruction to Municipal Officials from Mayor Kollek,
> 3 September 1967, quoted in Benvenisti 1976: 130)

Since Arab Jerusalem was incorporated into Israel, unwilling Arab citizens found they had the vote in Israeli elections. The first were the municipal elections of 1969. There was heavy Arab participation (although no Arab would stand) which helped to keep the Israeli Labour Party in charge of the city council, but it was participation through fear; rumours were rife that there would be no jobs for those who failed to vote. The new local authority co-opted leading Arabs on to municipal committees, but they all refused to serve. The initial co-operation had come to an end. Most leading Arabs now refused to co-operate with Israel, in some cases on principle but in other cases out of fear of reprisals from Palestinian nationalists. The Arabs could have had an important voice in a unified Jerusalem, but instead they lost all political influence in the hope of longer-term political gain. By 1983, Arab participation in municipal elections had fallen to 20 per cent and, by 1988, Arabs in east Jerusalem were rioting against the Israeli administration as part of the general West Bank uprising, aimed at the establishment of a Palestinian state including Arab Jerusalem (Benvenisti 1976: 139–48).

If a boundary could be removed by passing laws, Jerusalem would soon have been united by the Israeli *Knesset*. The Law of Unification in 1967, passed straight after the conquest of Arab Jerusalem, enacted a sudden and total change of administration. It was not thought out and was soon ignored and in disrepute. The law meant,

149

for example, that no Arab professional had the necessary licence to practise, and no Arab company was registered to do business.

A year later, the *Knesset* passed the Law of Legal and Administrative Arrangements. This was as arbitrary in some ways as the Law of Unification. For instance, Jerusalem Arab landlords were automatically registered in Israel and Jerusalem and Arab lawyers were automatically allowed to practise in Israel. Gradually, over the next ten years, laws were passed to deal with all the complexities of the economic as well as political relationship between Israel and Arab Jerusalem, and between Arab Jerusalem and the West Bank of which Jerusalem had been the principal city; this latter relationship was full of difficult problems, for example over the status of landlords outside the city boundaries in occupied territory and over the status of companies with the majority of shares held by people living in the occupied territory on the West Bank. Additionally, there was special legislation to deal with Mount Moriah, the ancient site of the Temple and holy for both Jews and Moslems who had built two important mosques there, Al Aqsa and the Mosque of Omar (the Dome on the Rock). A compromise was reached which gave Moslems permanent free access to the Mount, and exclusive access on Fridays and other Moslem holy days, to worship freely even at the Mosque of Omar on the very site of the ancient Temple's Holy of Holies itself, and gave Jews free access at all times to the Western Wall of the Second Temple, the sacred shrine of the Wailing Wall (Benvenisti 1976: 149–61; Romann 1981).

The most effective unity occurred where the law never reached. An undiscriminating Arab–Israeli underworld of drug barons, fences, and pimps soon developed profitably. A confusion of laws and authorities between Israel, Arab Jerusalem, the West Bank, and Jordan also offered unrivalled opportunities for financial crime.

What prevented effective unity at other levels was the simple fact of Israeli dominance. In the end, the power to impose was exclusively Israeli. Identity permits were forced on all Arabs; these were only obtainable by people who had been registered in a hurried and inaccurate census; yet not to have a permit could lead to arrest. Travel permits had to be obtained from Israeli authorities by Arabs who wanted to visit Jordan, and trade with Jordan was suddenly completely forbidden. Bus and taxi regulations now came under Israeli law, and pressure by the Israeli firms won them the right to operate in Arab Jerusalem, but Arab firms could not operate normally in Israel. Even the spread of Israeli trade unions to Arab Jerusalem, which local union officials saw as an extension of class struggle, came to be used as a means of weakening Arab business and diverting the Arab labour force to work in Israel.

The case of education was worse. A clumsy attempt to introduce quickly the curriculum used by Arab citizens of pre-1967 Israel and to censor or forbid a good deal of traditional teaching material caused a strong reaction and led to strikes in Arab schools and universities. It also led to a large-scale shift to private schools which are still attended by two-thirds of the children of Arab Jerusalem, even though the authorities have long since relented and allowed the Jordanian syllabus to be taught, plus Hebrew and civics (Benvenisti 1976: 195–205).

On top of all the problems of Israeli dominance and Arab resistance, there were the special difficulties in trying to unite what was basically a Third World Arab city with a developed, westernized Israeli city. For example, prices of fruit and vegetables, leather goods, and car repairs in Arab Jerusalem were threateningly low as far as the Israelis were concerned; firms went bankrupt as competition in some areas was impossible. Israeli workers were threatened by cheap Arab labour with no trade unions and Arab businessmen were stymied for capital as their Jordanian investments lost their value and working capital from Jordanian banks was inaccessible. Modern water supplies, sewers, and electricity were extended to most of Arab Jerusalem in a few years, but at substantial cost met by huge increases in local taxes which the Arab population were not used to paying. Generous Israeli social benefits were also paid for out-of-state taxation but this was still resented by the Arabs because of the large amounts spent on the army and police.

Physical unification

It has been through physical development that real reunification has been achieved in Jerusalem; the physical partition has simply been blurred. Physical reunification has taken the form of Jewish reoccupation of areas which were formerly Jewish; the clumsy Mamillah development proposal; infrastructural change; and the construction of major new Jewish suburbs in Arab Jerusalem (Prittie 1981, ch. 7).

The reoccupation of the Jewish quarter of the old city of Jerusalem started almost immediately after the 1967 war. The area had been neglected and in places desecrated by the Jordanians. At first sight, the boundary between Jew and Arab was merely shifted by this population movement; but the confined spaces of the old city and the need to deal with Arab landlords and official organizations made it inevitable that there was some rubbing of shoulders. This was a revival of the sort of formal relationship there had been before 1947.

The tasteful redevelopment of the Jewish quarter of the old city was in contrast to the extravagant proposals for the Mamillah area, which had been in the demilitarized no man's land between the Arab

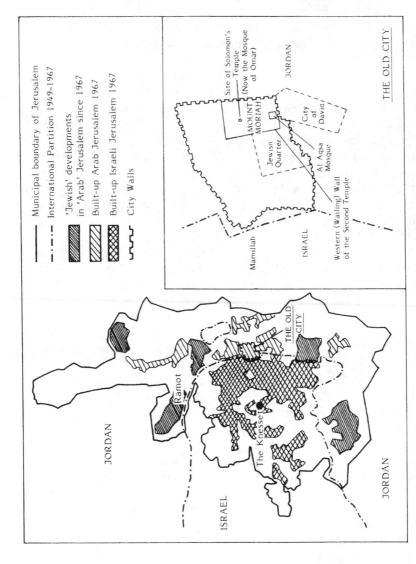

## Legend

- Municipal boundary of Jerusalem
- International Partition 1949–1967
- 'Jewish' developments in 'Arab' Jerusalem since 1967
- Built-up Arab Jerusalem 1967
- Built-up Israeli Jerusalem 1967
- City Walls

### The Old City detail

- Site of Solomon's Temple (Now the Mosque of Omar)
- MOUNT MORIAH
- JORDAN
- Jewish Quarter
- Al Aqsa Mosque
- City of David
- Mamillah
- ISRAEL
- Western (Wailing) Wall of the Second Temple

THE OLD CITY

### Main map

- JORDAN
- ISRAEL
- Ramot
- The Knesset
- THE OLD CITY
- JORDAN

*Figure 4.4* Partitioned Jerusalem, 1949–67 (also showing Jewish developments in Arab Jerusalem since 1967)

quarter of the old city and the main Israeli commercial centre. The Mamillah plan called in 1968 for a grandiose office and shopping complex on the site of a neglected Moslem cemetery. The project was discontinued for nearly twenty years, because it was considered architecturally poor and offensive to the Arabs. In 1986 a more modest scheme was started, which avoided most of the cemetery but still providing a physical link across the old dividing line.

Some of the most important direct physical links have been through the new infrastructure. Electricity, for example, forced new Jewish homes built beyond the old partition on to the Arab electric grid to avoid immense extra costs. New Jewish settlers and Arab electricity companies soon got used to dealing with each other on a day-to-day basis. The same applied to road repairs, street lights, and parking restrictions; in these sorts of cases, if life is to be bearable, co-operation is the only answer.

The life-style of Jew and Arab are very different:

Jews live in neighbourhoods of neatly constructed apartment blocks, while Arabs prefer large houses capable of accommodating the extended family. According to Amnon Niv, the municipality's chief planning officer, this trend may change. 'Just as Jews are more happy to live in houses if they can afford them,' he explains, 'so less wealthy Arab families will come to realize that it would be to their economic advantage to live in blocks of flats'.

(Griver 1986: 25)

The Arab taste for plenty of space has contributed to the fact that few extra houses for Arabs have been built since the Israeli occupation. Low density has provided an excuse for city planners to refuse building permits to Arabs developers who, in any case, find it almost impossible to raise the necessary finance. By contrast, 800 acres of expropriated Arab land is being used for flats for Israeli Jews. This is the most obvious physical expression of the reunification of Jerusalem. There is no ethnic or political boundary across the city. Although, at the most local level of the neighbourhood, Jews and Arabs remain separate, the physical arrangement before reunification when one half of the city was Jewish and the other half Arab has ended. The ethnic barrier along a physical line has gone. More ambitious developments have long been planned (Kutcher 1973) but for demographic reasons these are now grinding to a halt. Few Jews are now coming to live in Jerusalem; they feel threatened by Arab rioters and ultra-Orthodox Jewish zealots; Jerusalem lacks the cosmopolitan atmosphere of other Israeli cities and there are very limited job opportunities. The Jewish population of Jerusalem is shrinking. 'We are becoming a city of government employees and intellectuals,' Raif Davara, special advisor to the mayor, complained recently (*The Economist* 1987).

## Conclusion

Jerusalem was only partitioned by an international boundary for eighteen years. The partition separated two hostile and apparently incompatible communities. Jewish Israelis and Moslem Arabs, a part of the western world and a part of the Third World. Perhaps it was remarkable that for so long there were so few problems. There was always immense suspicion. When the Al Aqsa Mosque on Mount Moriah was burned by a deranged Australian Christian, Arabs accused the Israelis of cutting off water supplies and pouring petrol on the flames (whereas actually they called in fire brigades from miles around and rebuilt the Mosque out of public funds). On the other side, Israeli authorities saw large Arab families as a constant demographic threat and tried to set a 30 per cent limit on the Arab population of Jerusalem.

In the end, the nature of Jerusalem, including its unity or partition, depends now as it always has done from King David through the Crusades to the establishment of modern Israel, on politics beyond the control of any of its people. The Palestinian uprising of 1988, the Israeli political and military response, and the rise of both Islamic and Jewish fundamentalism are all beyond the control of the people of Jerusalem. Their city may divide again or unite more completely. Some of its population may be driven away or a mosaic of mutually tolerant minorities may live with each other in peace, as they did in the past, even though – now as then – they may never love each other:

> Jerusalem the Golden,
> With milk and honey blest,
> Beneath thy contemplation
> Sink heart and voice oppressed;
> I know not, O I know not,
> What joys await us there,
> What radiancy of glory,
> What bliss beyond compare!

> (Quoted in Grindea 1982: 213)

**Power through legality: summary table**

*Philosophy of power through legality*

1. Dominant values (modified by structural conditions and events) lead to the 'primary' objective in a *conflict over a legal matter such as a boundary.*
2. The 'primary' objective (modified by structural conditions and events) leads to the 'situational' objective in the conflict.
3. Political conflict theory triggers 1 and 2, above, into action.

The STATE BOUNDARY, the prime legal focus of conflict, varies in its significance according to its *delimitation* and *administration.*

Case-studies relating GEOGRAPHY to POWER THROUGH LEGALITY

1. SINO–SOVIET BOUNDARY DISPUTE
   A conflict of dominant values leads to conflicts of 'primary' and 'situational' objectives, and to limited conflict.
2. A FEDERAL BOUNDARY : BRITAIN–CHANNEL ISLANDS
   The same dominant values mean shared objectives and a range of mutually beneficial co-operation.
3. WESTERN SAHARA
   An attempt to construct a new state within former colonial boundaries.
4. END OF THE PARTITION OF JERUSALEM
   The enduring effects of a boundary through the heart of the Holy City and the efforts at overcoming them.

*Power through legality is determined above all by the international boundary, a legal entity of which the delimitation and administration profoundly affect the lives of people living in the states whose territories it defines.*

# Chapter five

# Power through legitimacy

## Introduction

Whatever the legal and constitutional position on the distribution of power, the truth is that it is distributed mostly to an elite who control or have influence over the rest of the population; in other words, elite power is hegemonic. Both within the state and within the community of states, the 'establishment' of capitalists, land-owners, and managers exerts a great deal of power over the rest of the population. So the real distribution of power is exceedingly unequal yet it is accepted as a framework for cultural and economic activity, sometimes reluctantly and sometimes willingly, by nearly everyone within the state and by virtually the whole community of states. This acceptance gives elitism legitimacy. Legitimated power enters every aspect of economic and cultural life; it is hegemonic in its control; hegemony is the consequence of power through legiti-macy, for hegemony would fail without legitimacy.

The first part of the chapter introduces the philosophy of power through legitimacy. It starts with Hegel's ideas (Hegel 1942) about those who are entrusted with the *Geist* or spirit of the state and goes on to consider Gramsci's ideas (Gramsci 1971) of pervasive bourgeois domination and the problems of opposing it. It concludes with Galbraith's insight (Galbraith 1967) that one elite is only in power so long as it controls scarce resources – once this elite was landowners, then capitalists, and now managers; theirs are the possessions or skills needed and they can demand the legitimation of their power as a return.

The second part of the chapter is concerned with the operation of hegemony on a world scale: 'Neo-imperialism: the legitimation of the developed world'. Hegemony does not work simply within the state but also between the states of the economic core (the advanced capitalist world) and the states of the economic periphery (the undeveloped world). The peripheral states serve the core states by producing primary products cheaply and they are dependent on investment and price determination by the core. The best example is 'agribusiness' which employs more than half the world's working population.

The case-studies are examples of the four key themes of power through legitimacy. The first, 'Ethiopia', illustrates neo-imperialism in action; the Ethiopian famines are shown to be, in part, the result of hegemonic relations between core and periphery. The second case-study looks at legitimated power within the states in 'Hegemony and Town Planning'. This explains how planners have always acted as agents of the elite or establishment of the state, and examines the insights and alternative ideas in Sennett's innovatory work. The case-study of Hip Hop looks specifically at cultural hegemony in western inner cities. Hip Hop culture plays an important part in providing a clear identity for young blacks, especially in the British inner city; it has not yet been possible for the cultural establishment, whose hegemony it threatens, to destroy it by incorporation. Finally, the case-study of Albania shows that hegemony does not need capitalism to function well; specific policies of hegemony for the ruling party in Albania have been highly effective in both the economic and cultural fields.

Overall, this chapter illustrates the effects of hegemony, the pervasive power of elites, legitimated by tacit consent, both within and between states. The use of this power through legitimacy has a profound effect on cultural and economic geography from tropical droughts to western inner cities; the geographies of Brixton or Ethiopia are frequently geographies of hegemony.

## The philosophy of power through legitimacy

### Introduction

The philosophy of power through legitimacy is the philosophy of the 'legitimated', or popularly accepted, domination of one group over others. Hegel, Gramsci, and Galbraith have each provided insights into how this kind of accepted power is achieved and wielded by the rulers of states. The dominant power of a ruling class over the rest of the population in a state is the hallmark of the 'hegemonic' state.

### Hegel

The idea of the hegemonic state started with Hegel (Avineri 1972; Hegel 1942). States had a spirit or *Geist* that was over and above the sum of the parts of the state; the *Geist* reached every part of the state. The ruling class was filled with the *Geist*; the rulers represented the spirit of the state. The acceptance by the rest of the population of the state spirit (including the idea of patriotic devotion

to the state), was therefore one and the same thing as the acceptance
by the rest of the population of the ideas of the ruling class.

Hegel then went on to suggest that, whatever may from time to
time appear to be the case, the ruling class must always inevitably
act in its own interest: 'Whatever is rational is actual and whatever
is actual is rational' (Hegel 1942: 10). A distinct class or group who
had a shared spirit and some ability to guide and rule always acted
rationally to maintain their own strength and power. This may or
may not be in the interests of all the rest of the population of the
state, but it must, by definition, be in the interests of the state as
a whole which was more than the sum of its people.

Hegel considered it inevitable that in the end any ruling classes
substantial enough to possess the state spirit should form a state. For
example, Germany, Italy, and Poland had failed to become states in
Hegel's day even though their ruling classes shared a communal
spirit or *Geist*. This was for particular historical reasons, but in the
end it was inevitable that they would be single states.

This state *Geist* excluded a good many of the people who lived in
the state. In his early work, Hegel specifically distinguished the *Volk*
from the rulers filled with the *Geist* of the state. The mass of the
*Volk* were the make-weights of the state, excluded from the closed
system of state power. It was a very short step from that system to
Marx's ruling bourgeoisie. But Hegel did not believe in a Marxist
revolution. Far from its inevitable downfall, he felt that a modern
ruling class could cope with the tensions of a modernizing society.
The style would change, as Hegel could see in his day (he lived
from 1770 to 1831) as the class of the monarchy, court, and landed
aristocracy metamorphosed into the class of the bourgeoisie, parliament, and civil service. The rulers acting rationally in their own
interests would know how to avoid revolution by mitigating the
worst social evils and by filling the heads of the *Volk* with
nationalism or patriotic ideals. It would be the dawn of the era of
the bourgeois hegemonic state with its control over the welfare and
thinking of the proletariat. Hegel considered the hegemonic state to
be natural, effective, inevitable, and completely resistant to
fundamental revolutionary change.

### Gramsci

Marx started with a much clearer definition than Hegel of the ruling
class. It was bourgeois capitalist, owning and controlling the means
of production. The state was merely a tool of the bourgeoisie and had
no special permanent characteristics or spirit; it would disappear
when the economic crisis of capitalism became so severe that a proletarian revolution was inevitable. The peculiarly Marxist interpretation

of the hegemonic state was developed by Antonio Gramsci in Italy after the First World War, when an explanation was needed for the lack of a successful world-wide revolution (Fiori 1970; Gramsci 1971; Simon 1982).

Gramsci's explanation of the durability of the bourgeois state centred on its intellectual domination. The bourgeoisie created intellectuals specializing in bourgeois economics and bourgeois literature to fool or comfort the proletariat respectively. Intellectuals were also used to run the organization of the state; they were the experts and the state's functionaries.

Gramsci argued that everyone was an intellectual and everyone was a philosopher. It was just that only certain types of intellect and certain philosophical ideas were recognized by bourgeois society. In fact, everyone's activity was technical, requiring learning, analysis, and understanding, and everyone had views, seldom expressed in bourgeois society, of art, the purpose of life, moral conduct, modes of thought – everyone indeed was a philosopher. There was a proletarian equivalent reality to the illusions of bourgeois class intellectual activity. Society was made up of bourgeois values and also of proletarian values; it was just that bourgeois values dominated in the bourgeois hegemonic state.

This imposition of ideas and values could only work superficially. The proletariat would accept bourgeois moral, religious, and social values at a certain scholastic level, but there was often a distinction between actual behaviour and considered scholastic statements about behaviour; hypocrisy was inevitable. The proletariat in fact had its own conception of the world which existed most of the time as a kind of subculture, only occasionally manifesting itself in action. Most of the time the bourgeoisie in the hegemonic state kept proletarian culture and philosophy submissive, subordinate, and out of sight.

In Gramsci's Italy, it was the Roman Catholic Church which provided the best example of the intellectual reinforcement of the bourgeois state. In the Roman Catholic Church, Gramsci argued, scholarly activity and intellectual debate were only allowed at the very highest levels of the hierarchy. Iron rules were imposed on everybody else. The educational institutions and the mass media imposed the acceptance of these rules through repetition (which Gramsci considered the best pedagogic means of working on the popular mentality) and through technical education (which would produce highly educated technical elites uncritical of the values of the hegemonic state).

There is then a contradiction within the state between bourgeois and proletarian values and philosophies, and there are different opinions as to whether this will produce confusion, passivity, or revolution. In the end, will the bourgeois state be reinforced or over-thrown? The Hegel school says that it will be reinforced, whereas the Marxist school says that deepening economic crises mean that in the end it must be overthrown. Gramsci drew the practical political conclusions that Communists, seeking the proletarian overthrown of bourgeois hegemony, should devote their energy to educating the proletariat until they have greater self-confidence in their own ideas and are willing to fight imposed bourgeois ideas. Communists would be working with the grain because bourgeois ideas could only ever be superficially imposed. Eventually the hegemonic state would be overthrown, but Gramsci never underestimated the Communist Party's task:

> [the Communist Party] must be, and cannot fail to be, the
> protagonist and organizer of intellectual and moral reform – that
> is, it proposes the terrain for a further development of the
> collective, all-embracing form of modern civilization. These two
> fundamental points: formation of a collective national popular will
> and intellectual and moral reform should constitute the structure
> of its work.
>
> (Quoted in Fiori 1970: 245)

### Galbraith

Gramsci saw the managers of capital as mere technicians in the hands of the true bourgeois hegemonists. Later in the twentieth century Galbraith took a closer look at the technicians and decided that they were as much a part of the ruling class as the owners of capital themselves. They were the 'technostructure' of Galbraith's *New Industrial State* (1967).

Galbraith accepted the basic Marxist tenet that power lay with capital but, he stressed, not with the ownership of capital as much as with its control. The power of shareholders was not that great; it was managers who had the real power, and they might own only a negligible amount.

Galbraith traced the history of the relationship between capital and power, starting in the early stages of the development of capitalism. In the seventeenth century no one doubted that power lay with the land. Land was scarce and accounted for perhaps three-quarters of all production. Landowners were the main users of capital and main sources of capital. Land was toppled from its position of power by mechanical inventions in the industrial revolution. Now there were factories, mines, shipyards, railways, and every kind of industrial development competing for capital. Some capital was still of course

160

used up on the land, but there were so many new uses for capital that capital became the scarce commodity and power shifted from the owners of land to the owners of capital. This was the age when businessmen and bankers made up the British and American cabinets. The United States Senate was filled with rich businessmen and the Chamberlains, screw manufacturers of Birmingham, rose to political prominence in Britain. Scarcity was power. Previously land had been scarce; now capital was scarce.

The power of capital was also destined to wane. Keynes had noted the tendency in modern society for surplus savings to accumulate. Capital is not normally scarce; it was scarce in the hundred years following the industrial revolution because the calls upon it multiplied so rapidly, but it would not always be so. In modern times, there had been a relatively high likelihood that business enter-prises can find all the capital they need. What has become scarce has been the ability to plan and use the available capital to make a good profit. The greatest need has become the need for specialized talent for organization, technical planning, and financial planning. Power has shifted to the manager who uses capital to maximize profitable return, and has little to do with the supply of capital. Capitalists and landowners have stayed rich, but it is the managers of capital, the 'technostructure', who have acquired the power because they control the commodity which is now scarce.

The operation of the hegemonic state ensures that the power of the people in charge of scarce resources is legitimated by popular support. Landowners, capitalists, and managers have each found in turn that power in the hegemonic state is delivered into their hands, as what they can supply becomes scarce and valuable. It ensures that their power is real power because it entails a control of the values and ideology of the state itself; their power is made legitimate in this way. This concentration of power has important geographical conse-quences as the case-studies in this chapter show. The group in power dictates the dominant value-system. Now, more than ever before, with the power of modern communications and the sophistication of marketing they can dominate effectively. At the same time, the increased accessibility of information technology and modern systems of communications may yet make a challenge possible from those who are ignored.

*Neo-imperialism: the legitimation of the power of the developed world*

### The definition and development of neo-imperialism

The equivalent of hegemony within the state at a world-wide level is the idea of 'neo-imperialism'. The idea of neo-imperialism is that developed and underdeveloped states are unequal but complementary parts of the same system. The underdeveloped world serves the developed world, which constitutes 'imperialism' in Marxist terms but, to distinguish it from the old empires, it is useful to call it 'neo-imperialism'.

The Marxists' view is that the system of neo-imperialism would develop in three stages. First, competition between capitalist firms in advanced states would lead to a constant pressure for expansion into exploitation of new natural resources, facilitated by improved transport and communications; at the same time it would lead to a constant search for cheap labour. Second, the searches for cheap labour and new natural resources would lead to a system of world capitalism in which the 'core' exploits the 'periphery' by direct extraction of profit and through monopolistic control of trade and prices. Third, within this world system, different forms of 'labour control' would be needed, ranging from forced labour to labour exploited by lack of alternatives through local elites (Brewer 1980).

Samir Amin took Marx's idea of the development of neo-imperialism somewhat further. Amin agreed that the capitalist processes are one single process involving the whole world. According to their role in the world capitalist system, states could be characterized as belonging to the world core or to the world periphery. Amin described the mechanism for the start and maintenance of exploitative core–periphery relations. He said that international specialization is determined by absolute cost levels and not by comparative advantage. Because the capitalist world must be seen as one system, comparative advantage (which only applies between economically independent states) becomes irrelevant.

Absolute cost levels depend on productivity and on wages. The states of the core developed capitalism earlier, and got a huge lead in productivity while wages there were still low. Later, wages started to rise at the core, but productivity was so far advanced in most cases that it continued still to be great enough for overall lowest costs to be at the core. Thus unequal development at the start turned out to be self-maintaining (Amin 1976; 1977).

The result for underdeveloped states was to give them a shared set of characteristics which may be summarized as follows:

1. Only a small proportion of the population is in modern industry.

2. There is permanent large-scale underemployment.

3. Agriculture suffers from low productivity.

4. Incomes are low except for a small elite; labour is cheap.

5. Agriculture consists of small peasant holdings side by side with modern export plantations.

6. Foreign trade accounts for a large proportion of production.

7. Capital goods are nearly all imported.

8. Raw materials are the most important exports.

9. Export earnings have to finance an outflow of dividends, interest payments, and royalties.

10. Trade with developed states predominates over trade with other underdeveloped states.

11. The small scale of industry and its domination by foreign firms leads to minimal development of bourgeois and proletarian classes.

12. Whole capitalist sectors are missing, especially sectors producing capital goods.

13. Modern export sectors operated by foreign capital operate side by side with primitive pre-capitalist sectors.

14. There is an over-expanded service sector as a result of a lack of investment opportunities in manufacturing.

The result for trade of this world system of neo-imperialism was that goods and capital flows within the core grew ever more important compared to trade with underdeveloped states. More and more accumulation of capital at the core, and therefore more and more wealth at the core, has led to trade within the core predominating over trade between the core and periphery. Trade with the periphery has become relatively marginal to the core, even though it is large in absolute amounts and crucial in its importance to the periphery. Evidence for this would be the successful development of the European Economic Community compared to the decline of trade between European states and their former colonies.

The central point of this world-wide capitalist system is an economic process: the effective use of capital in the periphery is blocked by the greater competitive strength of the industries of the core; the periphery's low wages cannot compete with the core's greater productivity per head. This greater competitive strength means the core can undercut the periphery and so make most manufacturing there unprofitable. This leads to the coexistence of pre-capitalist agriculture and urban shanty towns with modern foreign-owned exploitative industry (mostly in extraction or specifically requiring cheap labour) and agriculture. Coercion may be needed from time to time to ensure the continuance of the system, which leads to political support for oppressive regimes. This is the world system of neo-imperialism, a system of domination by the legitimated power of the developed world.

Amin ignores the state. He overlooks the way the state facilitates exploitation by giving national groups the illusion of independence and the determination to go it alone unprotected. If too much attention is paid to the state, however, economic failure is all too easily seen as a problem of state policy from which the state can therefore escape. The insight and relevance of this world view can be illustrated with reference to world-wide commercial agriculture, so-called 'agribusiness' (Frank 1979).

Agribusiness: neo-imperialism in action

Agribusiness is undoubtedly the most important economic enterprise in the world, employing over 60 per cent of the world's economically active population. This includes farm suppliers, farmers, processors, distributors, transporters, storers, financiers, lawyers, and salesmen, and covers all foodstuffs and fibres. Thus the whole massive system of agriculture is interrelated; this is 'agribusiness'.

Firms in agribusiness are not necessarily specialists, although the agricultural giants like Nestlé and Unilever are very important. Volkswagen grazes cattle in Brazil and Renault sells coffee-processing machinery to Morocco, while all the major banks are involved in one way or another. Of the hundred largest agribusiness firms, half are in the United States and the rest in Western Europe.

A topology of agribusiness firms' operations would include:

1. The direct purchase of land and operation of farms, especially plantations.
2. The purchase through intermediaries, or renting, of land, especially where direct purchase is politically difficult.
3. The opening of virgin lands, for example, in Brazil.
4. The supply of credit to local producers to buy imported machinery, seeds, fertilizers, pesticides, and other inputs from agribusiness firms. (Typically, these farmers will also deliver all or part of their harvest to the same firms or their associates for processing. The added advantage of this method for agribusiness is that the firms can easily pull out for economic or political reasons.)
5. Arrangements between the contractors, local landowners, and agribusiness firms for large-scale projects like dams or new roads. (These arrangements typically deal with the commercialization of agriculture made practicable by the projects. The catalysts for the

arrangements are usually either a bank with surpluses to lend or an international agency.)

The Sahel, the swathe of Africa bordering on the south of the Sahara Desert, is a good example of agribusiness at work under conditions of neo-imperialism. In the Sahel there had been traditionally and typically a ten-year minor drought cycle and a thirty-year major drought cycle. Then a drought in the mid-1970s led to terrible famine and, by 1977, the death of at least a quarter of a million people and between 30 and 40 per cent of the cattle. The reasons for the seriousness of that drought and subsequent droughts was intensive commercial farming for export and consequent deforestation and over-grazing. In other words, the cause was not the weather but agribusiness for export.

The socio-economic consequence of the famine was that land was sold off in substantial quantities, becoming concentrated in fewer hands. This led in turn to the devotion of yet more land to commercial as opposed to subsistence crops, and to worse droughts in the 1980s. Yet, at the height of the famine, agribusiness was making substantial profits from exports of food from Niger, one of the worst-hit states. In this system, food is sold according to economic demand and not according to need.

The underdeveloped world lacks bargaining power. Neither governments nor capital nor labour can get favourable terms and conditions from multinational agribusiness for the quality, suitability, and prices of the agricultural goods they import. Nor can they get the best prices for their exported agricultural produce; indeed local suppliers usually have to absorb most of any cyclical downturn in price. Worst of all, they can do nothing if, for example, soil fertility declines or a different government comes into office and agribusiness pulls out.

The few locally distributed profits of agribusiness are usually highly concentrated in the hands of a few landowners, producers, and merchants. The losses tend to be more widely distributed. Competing small producers, peasants, tenants, sharecroppers, and agricultural labourers may be displaced. The scarce supply of land, water, and natural resources, all of which have been directed to production for export, leads to shortages and high prices; so in the event of even a minor downturn either the poorest will starve or the state must spend foreign currency on subsidies for imported food. The land may be damaged ecologically, and then agribusiness will move on leaving calamity behind. On top of all this, rural unemployment arising from farming for export with technologically advanced methods leads to rural–urban migration and mass unemployment amongst slum-dwellers of rural origin.

Borrowing to pay for imported agricultural products; payments for licences, dividends, and know-how; and interest on loans from the World Bank and elsewhere leads to increasing direct economic dependency on the developed world. This loss of economic sovereignty leads to dependence on capitalist power to avoid financial collapse. Dependent states must therefore comply politically. This loss of political sovereignty is the ultimate consequence of the inequalities of power in a neo-imperialism world.

*Case-study 5.1: Ethiopia*

The development of the food crisis in Ethiopia

The food crisis in Africa south of the Sahara has been getting worse for over two decades. During this time a good deal of attention has been directed towards drought as a natural problem. Of course natural conditions are not unimportant but they are secondary to the international economic system of which Ethiopia is inevitably a part. Natural drought, after all, has occurred in Ethiopia throughout history without anything like the famines of recent years: even if the substantial recent increase in the population is allowed for, a further economic explanation of the food crisis is needed.

Ethiopia is one of the least developed states on the poorest continent in the world, and her economy has stagnated. For thirty years economic growth has lagged behind population growth. Agriculture accounts for about half of GNP and about 90 per cent of exports (mainly coffee and tropical fruit); it also accounts for about 90 per cent of employment. Ethiopia is therefore heavily dependent directly and indirectly on international agribusiness.

Four other international factors have made the economic conditions in Ethiopia even worse than might otherwise have been expected. First, the rise in oil prices in the early 1970s coincided with a revolution which brought to power a military government politically allied to the Soviet Union; although the Soviets provided plenty of military aid, they could not match the loss of economic aid from the west. Second, aid cuts particularly hit training, which meant that Ethiopia, perhaps more than any other state in Africa, lacked the skilled manpower to make the best use of whatever capital investment was available. Third, there was a major downturn in world coffee prices between 1979 and 1982 which cut Ethiopia's export earnings by nearly 30 per cent in three years. Fourth, boundary disputes combined with ideological differences to start an international war with the Somali Republic; although there is no longer open warfare, the diversion of military resources has provided an opportunity for separatist movements within Ethiopia to progress, especially those in Eritrea and Tigre (Figure 5.1).

One international factor which has never featured in Ethiopia's food crisis is a world-wide grain shortage. Across the world, grain stocks far exceed the shortfall for the whole of Africa. It is the strong international economic demand for cash crops that has created the grain shortage in Ethiopia. Almost all agricultural investment between 1960 and 1980, at the height of the 'Green Revolution', when Ethiopia could have been shifting towards self-sufficiency, was in the cash crop sector. The Green Revolution's fertilizers were scarcely even made available to the small farmer growing food.

### The case of agribusiness in the Awash Valley

The best conditions which Ethiopia could offer agribusiness were in the south. It was mostly pastoral land and the pastoralists could easily be evicted or cheaply employed to work on the new plantations – they certainly had no power to resist. The pick of the areas available was the Awash Valley which had the important advantage of being near to the relatively industrialized and cosmopolitan capital, Addis Ababa. The scheme for the Awash Valley to be taken over by a Dutch multinational in 1954 had the full backing of the government of Ethiopian Emperor Haile Selassie. Indeed, the government paved the way by evicting the Gile tribe of pastoralists (Halliday and Molyneux 1981: 65–7; Markakis and Ayele 1978: 77–81).

The aim was to produce tropical fruit for export and sugar to substitute for imports. The Dutch were prepared to go ahead only on guarantees from the government of a high tariff on imported sugar (creating a virtual monopoly for Awash Valley produce), of tax exemption for all goods or fuel imported by the company, and of an easily exploited labour force that required a company of government police stationed next to the plantations to keep the labour force in line. The multinational did well and the Gile community disintegrated and dispersed to destitution and to be among the first victims of the famines of the 1970s.

### The causes of famine in Ethiopia

Drought was the immediate cause of the first of the great famines but it was not the prime cause. The famines occurred not because there was not enough food in Ethiopia but because production was organized in the interests of international capitalism; this brought about a catastrophic decline in the purchasing power of a large number of Ethiopians: 'Starvation is the characteristic of some people not having enough food to eat and not the characteristic of

167

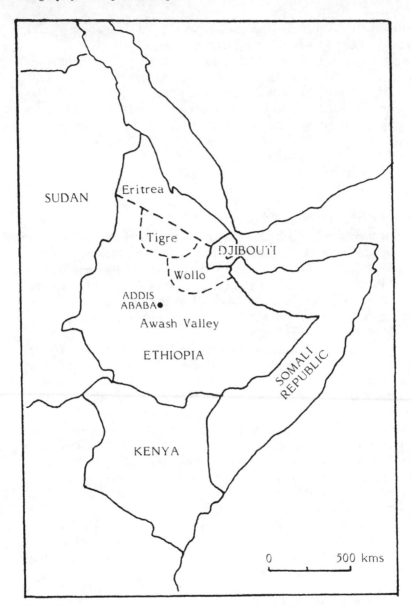

*Figure 5.1* Ethiopia

there being not enough food to eat' (Sen quoted in Snowdon 1985: 49).

What matters is not the need for food but either the ability to pay for food or the direct ownership of food. The peasant farmer must therefore either own land or sell his labour power for food or for money. If he owns land, he may grow cash crops and then use the money he earns to buy food. Whichever way it happens, the system can easily go wrong: the cash crop may lose its value on world markets; price controls on produce may mean that hired labour is sacked; or there may be crop failure due to disease, drought, or war. Nearly all the causes of disaster are economic or political; only disease and drought are natural.

When the first big famine of the 1970s afflicted the people of Ethiopia, food production declined by only about 7 per cent, mainly because of local crop failures in the central province of Wollo. The problem was that the commercialization of agriculture, the civil war, and the destruction of much of the infrastructure meant that there was no escape. In addition, food actually moved out of Wollo during the famine to meet economic demand (Snowdon 1985: 50). It was perfectly reasonable that whatever little food was produced amidst the cash crop plantation of Wollo should be taken away, because the purchasing power of people in Wollo had collapsed. Moving food for sale into such an area is pointless because no one can afford to buy it.

## Government responses to famine

The original Ethiopian government reaction to the famine was to deny it was ever happening. In Haile Selassie's empire there was a virtual conspiracy of silence (Halliday and Molyneux 1981: 67). Under the revolutionary government, there was a great reluctance to invite aid because of the political influence that could accompany it.

The next reaction by the Marxist government was to treat some of the symptoms of the famine. The natural tendency for the price of food, as it became scarce, was to rise sharply, but the Ethiopian government's policy, in part to retain support in the capital and other main centres, was to keep food prices down. Artificially low food prices imposed by the government were a further disincentive for farmers to grow the surplus needed and also led to the sacking of farm servants on a large enough scale to create a major problem of destitution by itself. These policies were certainly poorly thought out and appear to have been motivated primarily by a misapplication of ideology and a misjudgement of political self-interest by the regime (Hancock 1985).

The more recent reaction of the government has been to take direct control of agriculture by reorganizing the population and farms and bringing them into larger, manageable units in areas where their production could be planned centrally. Their longer-term aim is to create a balance of cash crops and food production for self-sufficiency and growth. The aim is to consolidate small and fragmented holdings, to promote the use of modern technology, and to innovate traditional crafts to the level of small-scale industry. The key seems to be to create units which are large enough to be capable of being planned but small enough to manage as co-operatives and so give their peasant members a say in how they are run (Chole 1987; Ghose 1985).

There are three stages of co-operative development envisaged. The first of these, the *Malba*, is common throughout Ethiopia. The second is the *Welba* which is still rare. The third is the *Weland* which has not yet been achieved anywhere. *Malba*, *Welba*, and *Weland* represent successive stages of co-operation and collectivization.

The *Malba* is an elementary producers' co-operative. In a *Malba*, the farmland is collectively operated, except for one-fifth of a hectare for each household left for private cultivation; members allow the use of their private farm implements and draft animals for co-operative production, and the distribution of income is in accordance with the labour and goods contributed.

The *Welba* is an advanced producers' co-operative. All the land in a *Welba* is collectively operated except for one-tenth of a hectare per household. All farm implements and draft animals are collective property, and income is distributed in accordance only with labour inputs, with allowances for age, disability, and so on.

The *Weland* will be a high-level co-operative with all the land and means of production of several *Welba* put together and brought into collective ownership. All *Welba* members involved will then become one centrally organized production brigade, and there will be no private production.

The co-operative system suits the traditions of the Ethiopian peasantry reasonably well. They have found it quite difficult to accustom themselves to being individual private farmers in a capitalist system which was the immediate result of the Ethiopian revolution's elimination of large domestic and foreign landlords. Indeed, this small-scale capitalism caused severe difficulties when the government intervened on prices and consequently discouraged food production. With these larger units, the Ethiopian government finds it much easier to institute schemes involving anything but minute amounts of capital than with numerous peasant holdings. For example, small-scale water conservation schemes, soil enhancement schemes, farm tool improvement schemes, and so on are all much easier and more effective in a more centralized co-operative farm system.

In some cases, however, a desperate shortage of foreign currency, arising in part from military spending and debt repayments at high rates of interest, has led to some co-operative farms at *Welba* level behaving in exactly the same way as the agribusiness multinationals in the Awash Valley years before. There has in these cases – and they account for up to a quarter of all new government investment – been no effort to promote food production and no effort to take on more labour to bring more people within the planned economic system. On the contrary, once co-operatives start to prosper from cash crops, they are not keen to attract new members to share their new-found wealth; some of them have mechanized production, incurring substantial debts and making themselves dependent on imported fuel, machinery, and spare parts and on outsiders for repair and maintenance services.

There is little doubt that in Ethiopian conditions, co-operative farming could be an important way of protecting people against the worst excesses of the neo-imperialist economic system. It is a way of running large enough units to facilitate the planning of production and prices and therefore quite an effective way of making use of limited capital to employ surplus labour in an organized way. Craft industries, crop variations, small-scale construction, and effective farm improvements have all made important strides in co-operatives in Ethiopia; this would not have been possible either with the reactionary landlords of the empire or among a mass of independent peasants. On the other hand, it is particularly difficult for the co-operative sector to develop successfully when continuous wars make the need for foreign currency to buy weapons an overriding imperative, because the co-operatives that should be producing grain for food must produce passion fruit and avocado pears for export.

The outlook is bad, but at least Ethiopia had devised a method of organizing production which combines some of the advantages of private capitalism with some of the advantages of socialist planning. If it is given a chance to develop, it may well be a model for other states locked into a system of cash crop production and food shortages.

*Case-study 5.2: hegemony and town planning*

Introduction

Planners play two main roles in the hegemonic state. First, their plans are likely to be favourable to the financial and commercial interests of owners of property. Second, they can play a key role in keeping society ordered and quiescent. These two roles are considered here in turn.

Planning for the establishment

The argument that planners act directly on the interests of the establishment starts with the observation that planners restrict the supply of land and thereby increase its price. Since land is privately owned, this must benefit the landowner at the expense of everyone else.

A good example of planners' restrictions boosting prices is the English Sussex coast. In Sussex, structure plans and local plans are highly restrictive. Housing is scarce in relation to demand and therefore very expensive, and many young people find the price of housing so exorbitant that they are forced to live in cramped conditions or to move away. One official reason advanced for the restrictions is the preservation of the countryside, but planning law allows farmers to do almost exactly what they like in terms of building and destruction of hedgerow and downland. A more plausible explanation is that local authority planners and the planning law they operate actually represent directly the interests of property owners, landowners, and estate agents, many of whom are also councillors and therefore also directly in charge of planning. Bramwell and Slowe (1986) have contrasted these establishment interests with those of most local people and the local economy in their paper 'The Costly First Home'.

A well-known example of planning in broader establishment interests is the raised section of the main western access to the centre of London, the A40 motorway. 'Get us out of this hell,' the banners cried from first floor flats, 20 metres from the traffic. Eventually people were moved out from those directly affected corners of Paddington and Praed Street, but their inner suburb was blighted for all time. Meanwhile the people from the wealthy western suburbs and stockbroker-belt villages in the Thames Valley found their journey into London had speeded up remarkably. It was also good for car manufacturers; more people would use their car and fewer would go by train.

In all these ways, planning restricts people's options. It can keep them quite literally confined, as Harvey illustrated in his study of the black American ghetto. Zoning regulations can make it too expensive for any but a tiny minority of blacks to get out of the ghetto areas in the centre of American cities. They remain in the ghettos, targets for exploitation. Landlords of the public and private sectors are free to charge them a 'class monopoly rent' (Harvey 1973: ch. 5). The overall aim of the establishment may be summed up as capital accumulation. This is assisted by specific planning decisions like those in Paddington and in West Sussex.

Planning for an ordered society

Capital accumulation is also made possible by the acceptance, in the hegemonic state, of the carefully planned city as a proper physical order and the basis of a proper social and economic order. The way this ordering works was brilliantly described by Richard Sennett in *The Uses of Disorder* (1973). 'The institutions of the affluent city are used to lock people into adolescence even when physically adult.' Sennett described the modern western city as a place of technical complexity but social simplicity with ever fewer interactions between different sorts of people, an unstimulating purified community.

Sennett's argument was that contact points had narrowed and that society had become duller, more controlled, certainly more controllable, and therefore much poorer. The upward movement in material wealth had been matched by social withdrawal. The older forms of complex association, related to cultural, class, and ethnic mixing in the inner city, had been replaced by a simpler kind of contact structure. This new pattern was embodied in the planned distinction between city and suburb which encouraged an artificial and undesirable ordering of society. Suburbanites found security and sanctuary through exclusionary measures based on race, religion, class, or other 'intrusions' on 'a nice community of homes'. The alternative to mixing was the growth of a strong and intense family life, which Sennett described in fascinating and horrifying detail. Intense suburban family life was in reality an admission that people no longer felt confident of their own ability to cope with a more mixed, richer environment, but only with the neat lawns and tidy supermarkets of suburbia. The argument was that people only had the confidence to live in isolated and easily understandable communities of similar people.

Sennett concluded that it was much easier to control a simple and organized city than something more complex. Haussmann's design for Paris (as well as providing boulevards suitable for the use of grapeshot) consciously planned physical space for predetermined social use. Instead of assuming that changes in the social structure of the city should be accomplished first in order to change the physical appearance of the city, Haussmann bequeathed the notion that it was simpler to change the physical landscape in order to alter the social pattern of a city.

Sennett's thesis was that, ever since Haussmann, hegemony and planning had gone consciously hand in hand. Planners laid down boundaries of social and economic areas as straight lines, and it was only a short step to the idea of people straightjacketed into suburbs where their lives were controlled and controllable. Planners were the

technicians who created and then preserved the physical characteristics of this sort of city.

*The alternative: planning by the public not professionals*
Where public 'participation' in planning in Britain and the United
States has been tried, it has only been in reality a disguise for
hegemonic control. For example, one London planner has
commented:

> public participation was little more than a belated information or
> last-minute consultation exercise. Decisions were effectively made
> before consultation took place with semi-secret deals between
> councillors and developers, or else central government overruled
> a decision taken democratically at a local level. The experience
> of much town centre redevelopment in the early 1970s is a case
> in point. Where public participation failed, it was because the
> politicians and the planning officers never wanted it to be real in
> the first place . . . for public participation to have any real
> impact, the politicians must want it, believe in it – and make it
> happen.
>
> (Colenutt 1988: 1)

Sennett's argument for a future of planning based on thorough
public participation was intriguing. He accepted the view that for any
alternative system of planning to work within the hegemonic state,
it must offer a better chance of order and security than the present
system. He argued that the planning of zoned neighbourhoods led to
the highlighting of racial and class differences with a greater risk of
violent conflict as a result. Mixed neighbourhoods have less violent
conflicts, although there is plenty of tension. In mixed neighbourhoods, because people have no alternative but to live together, most
tension is peacefully resolved. And there is no possibility of mythical
images of other classes or races developing.

Sennett advocated that Haussmann's assumption should be dropped
from planning. The planning of cities should no longer be concerned
with order and clarity. Instead, the city should be conceived as a
social order with no coherent forms. The planning of functional divisions and land use in advance of habitation, or to fossilize an order
already in place, should be abolished. City spaces should be for
changeable and varied use. There should be no zoning and no
control.

Sennett argued that the pre-planned image of class neighbourhoods
and functional neighbourhoods, definable on planners' maps,
promoted adolescent ideas in people's minds; people came to feel
that they belonged to pure communities of a particular type with a

clearly defined geography and certain proper values. Sennett argued that this was incompatible with the aims of the hegemonic state, because such pure order needed constant heavy policing which in the end had to fail as tensions inevitably grew with natural changes in society over the years. Still less was it in the interest of individuals because it was so dull and unstimulating.

Unplanned and unzoned areas should be encouraged, and there should be no more central control. This would promote visual and functional disorder in the city. It would be a revolution against the dead hand of predetermined planning which restricts social exploration, insists on a single culture, and dampens individual potential by restricting creative challenge and stimuli. The community free to create its own patterns of life would be very exciting. In the inner city, stockbrokers and labourers, immigrant clusters, and enterprising shopkeepers might live next to some light manufacturing, all-night bars, stores, and restaurants. There would be a high level of tension and unease, and the air would be full of conflict between distinctive groups of people. In Sennett's city, both the threat and the assurance of police control would be gone, for the police would not have the responsibility to keep peace in the community by repressing deviance but rather would be restricted to dealing mainly with organized crime. People would really be forced to deal with each other. Problems would have to be worked out and, because everyone would belong to the same neighbourhood, know about each other, and have no purified images of themselves or other groups or classes, most issues would be resolved without violence. There would be picketing and boycotting, meeting and organizing; above all, it would be a stimulating alternative to town planning. Doing away with this kind of town planning would probably be less of a threat to the hegemonic state than maintaining artificial polarization.

*Case-study 5.3: Hip Hop*

Hip Hop: a threat to cultural hegemony

Just as town planners dream of pure, comprehensible, and controllable physical communities, so the cultural establishment likes to categorize, understand, and control minority cultural movements – otherwise they might become a threat to cultural hegemony. Indeed, in the end, the existence of an alternative culture or life-style may threaten not just cultural but even economic and political hegemony. The concept of a working alternative is a threat to those with power. So Hip Hop culture is a threat; it is inaccessible to most people, associated with behaviour considered to be antisocial and, above all,

it has been adopted by black inner-city youths, one of the few groups
to riot and protest in Britain in the 1980s. The response has been
an attempt by the establishment to incorporate Hip Hop into mainstream culture through the major record companies; but this has been
only partially successful because Hip Hop is a broad culture, much
broader than could be contained on a record, with meaning only to
people in the inner city, especially to Afro-Caribbean black youths.
Black youth can also transform it subtly and continuously and, in
that way, resist incorporation by changing the culture itself. Additionally, the very threatening nature of Hip Hop is one of its main
selling points to the public beyond the inner city, so short-term
profitability depends on a long-term risk to cultural hegemony.

What is Hip Hop?

Hip Hop is a combination of cultural characteristics. Its main components are rap (rhyming or talking against a background of a
regular, loud, and persistent beat); scratching (the deliberate scratching of a record using a stylus to produce an additional clicking
beat); cutting and mixing (copying from a whole variety of different
sources to create one track); using words as melody (as opposed to
musical melody); graffiti (in a characteristic style giving the appearance of three-dimensional writing); breakdancing with its associated
body-language and postures; and certain kinds of clothes, particularly trainers and tee-shirts with graffiti prints. Fab Five Freddy
in the film *Wild Style* was the first to bring all these cultural
components together systematically, but the assemblage was in place
anyway in the Bronx in New York by the start of the 1980s and in
inner London by the mid-eighties.

Hip Hop had its origins among black youth living rough in New
York. The original records were taken from open-air park concerts
in the Bronx, laid on and policed by gangs which were often also
involved in drugs and racketeering. It happened that several of the
gang leaders were also entrepreneurs who saw a moneymaking
opportunity in Hip Hop. There was a particularly good market for
rap among less impoverished black youths and eventually among
well-to-do white students at New York University; then the big
record companies and the media moved in. KRS1, a leading rap
artist, describes the contrast between the marketed image and the
reality when discussing the rise to fame of the gang leader Afrika
Bambaataa ('Bam'):

> It's real funny when you think about it. You had to be criminally
> minded to get to the top . . . . I would see the media print
> beautiful things about those guys, but around my block it was

nothing to see them in the middle of the street with a sawed-off shotgun, cops passing them by, too scared to get out of their cars. In those days I was a graffiti artist and knew everybody through graffiti . . . the graffiti artists were ripping the city down in the train stations, and Bam and his Hip Hoppers were rhyming and robbing.

(Halasa 1988a: 23)

## Hip Hop in the cultural landscape of the inner city

Hip Hop came to Britain a bit at a time, starting with breakdancing. Breakdancing gave teenagers a sense of power; it was a ritual display of postures not easily comprehensible to outsiders, which gave them the chance to use to the full the strength and flexibility of the body at an age when the body stands out as the means of personal expression, particularly in circumstances where there was little money, poor schooling, and few jobs. The inability of outsiders to understand was especially attractive and hastened the introduction of other aspects of Hip Hop, including rap: the loud beat, like the breakdancing, was another opportunity for personal expression.

At first, it was second-hand Hip Hop that came to Britain – the rap, graffiti, and clothes were all the same as in New York – but black schoolchildren in Brixton and Notting Hill were soon writing rap rhymes themselves and adding from their own local experiences to Hip Hop culture. It was inevitable because Hip Hop depends heavily on its local environment. Unlike reggae it is not concerned with roots; instead, it is a collage of diverse aspects of current experience. Hip Hop appeals equally to the young, black, and poor in Brixton as in the Bronx, but their circumstances are different and so Hip Hop adapts differently.

Hip Hop is avowedly modern and unconcerned with roots. In Britain it is in some ways a reaction against the Afro-Caribbean rhythms of reggae. Everything about it is a metaphor of the world inhabited now by young blacks in the British inner city. Its use of record players rather than expensive musical instruments and its simple beat boxes (digital electronic drum machines) reflect the technology available to inner-city youth; the dislocated form of the cut, scratched, and mixed sounds reflects the nature of their community.

Black youths in inner-city Britain lack the sense of identity of their New York counterparts, but Hip Hop has provided a cultural framework which is different from others, because it demands a contribution from its adherents. Hip Hop music, graffiti, and clothes provide the space and framework for a message, but they do not

177

write the message. The strength of Hip Hop – and its threat to an establishment which prefers anonymous and disorganized inner city poverty – is the way it forces its adherents to think: 'Consumption is turned outwards: no longer a private, passive or individual process it becomes a procedure of collective affirmation and protest in which a new authentic public sphere is brought into being' (Gilroy 1987: 210).

### Hip Hop and black youth identity

Hip Hop has made black youths define for themselves their own cultural identity. There are five main aspects to the choice they have made. These are their acceptance of negative labelling, their expression of strength, their determined inaccessibility, their increasingly political choice of topic, and their geographical specificity.

First, any group will tend eventually to inhabit the label they are given. Black inner-city youth in Britain possesses mainly negative labels as far as the media and much of the rest of the population are concerned. It responds by fulfilling the stereotypes of criminality and by rejecting mainstream British culture. Some rap artists actually base their appeal on the stereotype of the black criminal and generate an atmosphere of violence and confrontation when they perform. A typical rap record label may feature angry youths by a burnt-out car.

Second, Hip Hop provides an opportunity for a ritual expression of strength. The body-language associated with it is derived from the codes of American gangs. The postures require discipline and imply machismo.

Third, aspects of Hip Hop culture are made intentionally inaccessible to most of the population. The sound is not easily intelligible to people who are used to musical melodies and at the same time its beat is obtrusive which makes its inaccessibility all the more significant. If it were quiet and inaccessible, like the Indian film music popular with the Asian British, then it would pose no threat. But it is not quiet; it is meant to be heard in the city, on the buses and tubes and on the pirate radio stations.

Fourth, the choice of the topic for rap is important because it is all the more clearly audible without music and is also impressively demarcated by a very strong beat. Sometimes the topics are mere gibberish, but they can also be very serious. Typically they are about the subjects of the street such as drugs, money, and sex, but they can also be overtly political such as Overlord X on Death Row and the Latin Rascals on Malcolm X. The location of performance gives added strength to a political or anti-establishment topic; typical venues are municipal halls or commercial warehouses. In *There Ain't No Black in the Union Jack*, Gilroy comments on the importance of location: 'The town halls and municipal buildings of the inner city in which dances are sometimes held are transformed by the power of these musics to disperse and suspend the temporal and spatial order of a dominant culture' (Gilroy 1987: 210).

Fifth, Hip Hop counters the anonymity of city life. It is associated with specific and definable black neighbourhoods. In the United States, Hip Hop is associated with gangs which are strictly territorial; graffiti and bandannas are the signs of the control of specific geographical areas by particular gangs. In Britain, ghettoization is far less complete, the black population is smaller and gangs are less sophisticated and less violent. But the readily identifiable character of Hip Hop allows black neighbourhoods to be defined as never before. Just as the state is reified by national territory and finds its identity through nationhood, so the black inner-city community is reified by cultural territory and finds its identity through culture.

### Hip Hop's struggle with the establishment

Hip Hop culture is strong enough and influential enough to be perceived as a threat by the cultural establishment. It has given a new aspect to life in the British inner city, and a new identity and strength to the black inhabitants. The only response by the establishment has come through the major record companies whose prime aim is to make a profit out of rap. They market rap by stressing the dangerous stereotypes of criminality, which reinforces the negative image of the Afro-Caribbean black British; in fact, the records are often marketed as though they came from another planet. At the same time, the topics are sometimes so trivialized that rap has even been used in an advertisement for Barclays Bank!

The abuse of rap make little difference to Hip Hop culture in Britain because there are so many other aspects of Hip Hop culture. Breakdancing may feature on 'Top of the Pops', and Reebok Shoes may profit from the fashion for trainers, but postures and fashions are too subtle to lose their identity just because some version of them can be found in the British mass market. The main risk to Hip Hop culture is that major record companies may damage rap by putting small independent companies out of business and then neglecting it. Certainly, if they should wish to damage rap, they are now in a powerful position to do so. They have signed up very large numbers of rap artists and are increasing their stake all the time.

Meanwhile the political temperature of Hip Hop is getting higher. The thrust is from the United States. Leading rappers, such as L. L. Cool J. and Schoolly D., have strongly political rhymes critical of the police and of capitalism, and the Black Muslim movement is gaining new supporters among Hip Hoppers in the United States and Britain. The public relations officer of one of Britain's and America's leading rap artists, Public Enemy, was quoted by Halasa as follows:

> All Public Enemy is trying to do is to raise the conscious level of black people, and we get so much grief from doing that. You should see the hate mail and threats that are sent to our studio, but it doesn't bother us. There are more pressing issues to be concerned with. The question is, who will speak out for the hurt masses of black people in America? Another woman was found dead, raped, and murdered and they engraved 'KKK' on her body. For me as a black man to see that on TV, to sit back and not feel something, I would be a fool. We don't get justice in America. We're constantly brutalized by police – in America you have to carry a gun to protect yourself from the people paid to protect you – we're being taken advantage of at our jobs, losing jobs, losing land, disrespected at work or in the school system. What are black people to do?
>
> (Halasa 1988b: 14)

This public relations man has been banned by Public Enemy's record company for his political remarks.

Public Enemy's public image illustrates the problem the cultural establishment have with Hip Hop. Their image is crude, wild, and dangerous. The record companies make money out of Public Enemy and hundreds of artists like them partly because of this image. Therefore, when Public Enemy decides to be political and more obviously threatening, the record companies (whose first and foremost aim is profit) cannot easily resist. Besides, it is not just rap but the whole broad range of Hip Hop culture and life-style which is providing, at last, a clear identity for the Afro-Caribbean youth of the British inner city. Cultural hegemony is in a serious dilemma.

*Case-study 5.4: Albania*

Introduction

Hegemony is not the exclusive preserve of the bourgeois state. Albania, an independent state about the size and population of Wales and situated on the Adriatic between Greece and Yugoslavia, claims to be a post-revolutionary society where the proletariat has taken control. Under these circumstances, the proletariat must struggle for hegemony in the face of a constant threat from the revival of bourgeois power. The aim in Albania was to use the organized strength of the Party of the proletariat as the main weapon in the early stages after the revolution and then to devolve power from the Party centrally to the proletariat – to the workers themselves. This second stage could only be risked after the Party's position of authority was thoroughly entrenched – after a thorough revolution of values and an unrelented programme of propaganda.

Albania was occupied by the Italians in 1939 and was of considerable value to the Axis powers in the Second World War because of its chrome and oil. But the infrastructure was destroyed in the war and the Communist partisans who took power on liberation inherited only the potential for economic independence on condition they could raise, without strings, the capital for rebuilding. The new government became isolated from the west over a dispute about British naval vessels sunk off the Albanian coast and was anyway suspicious of quasi-colonial intentions. The Soviet Union became Albania's backer in return for a naval base.

The leader of the (Communist) Party of Labour, Enver Hoxha, announced a Stalinist planned economy in a one-party state in 1947. He was severely constrained at that time because there was very little industry to plan and very few educated people to administer a plan. Nevertheless, all public utilities and capital owned by foreigners or Albanians were nationalized in 1947 and broad targets for production were agreed. By 1953, a Soviet-style planning system had generated such a massive bureaucracy that Hoxha was able to thin it (and purge it) by 30 per cent in a few months. Despite the purge, the Stalinist centralized system remained with its strictly hierarchical principle of *Udheheqjeunike* (one-person leadership) being applied right through the system; from national planning to enterprise planning, individual Communists with technical expertise were the official leaders. Enterprise directors were controlled by the ministries and the ministries were controlled by the Party and there was absolutely no worker participation.

Policies for economic hegemony

Albania broke with the Soviet Union over de-Stalinization in 1956 and the end of Soviet aid and training facilities left plans unfulfilled. Albania would soon need more foreign capital and at the same time was determined to avoid international debt with consequent political dependence (in fact a law was passed against accepting foreign credit precisely because of its political implications). China filled the aid gap to some extent but resources were more limited and supply lines complicated. By 1965 there was large-scale absenteeism and disguised unemployment, and an enormous cadre class had emerged with all the characteristics of a pre-revolutionary bourgeoisie; a wealthy middle class of planners and enterprise directors. There was

181

a contradiction between the emergence of this new class and the revolutionary aim of eliminating bourgeois characteristics. The contradiction was resolved by starting a major campaign to develop the hegemony of the Party of Labour.

Party officials were trained to be political commissars and to work alongside the labour force. The state plan and the enterprise plan were discussed alongside the aims of the revolution and Communist ideology. The purpose was to spread the proletarian *Partishmeria* or Party spirit as widely as possible so the Party could trust the commitment of more of the population to the aims and values of the revolution and rely on more people to work willingly to fulfil the targets of state and enterprise plans.

Workers were consulted and plans discussed but there was not yet any real decentralization of power.

> Responding to the call of the Party, the masses of the people
> discussed indices of the fourth Five-Year Plan in such a lively
> and creative manner, and with such a profound and revolutionary
> sense of responsibility, that they set themselves numerous indices
> exceeding even the most optimistic produced by the state.
>
> (Institute of Marxist-Leninist Studies 1971: 597)

On the whole, workers accepted all the Party's very demanding targets, partly because any individuals who objected risked being accused of treachery to the cause of the revolution.

Between 1967 and 1970, a great many cadres were sent to factories and farms and infiltrated the workforce. Then the Party leadership took the next step in the elimination of bourgeois characteristics and extending the Party's hegemony. They eliminated most of the power and privileges of enterprise directors and senior technical staff:

> If a director keeps making mistakes the workers can say: 'We
> will throw you overboard and bear well in mind that there is no
> one who can help you; the Party is ours, the regime is ours, it is
> we who are in power, it is the dictatorship of the proletariat
> which reigns.' The labouring masses should by all means and
> without hesitation strike down the director of this type or any
> other functionary of this kind, whoever and of whatever rank he
> may be in the Party or the government.
>
> (Hoxha 1975: 43)

This initiative opened the way for some real advance in workers' control. Inevitably there was sometimes chaos and sometimes a vendetta against management. There were also striking successes where fraud or gross inefficiency were uncovered and then put right with large increases in production.

The initiative was divided in practice into two stages. First, the Party's influence was increased through an intense programme of propaganda using every conceivable means of communication and through the infiltration of the workforce by Party officials reporting straight to Party headquarters, the planning ministries, and the secret police. Second, the aim was to get the workers to fulfil the production plans, rather than enterprise managers who could no longer be relied upon to be faithful to the Party because they were becoming a managerial class with bourgeois cultural characteristics and lacked shop-floor (or farm) experience. Ten years' experience of manual labour at a factory or farm was now demanded of all managers and administrators, and the salaries of officials, experts and bureaucrats were reduced to approximate parity with the rest of the workforce (Institute of Marxist-Leninist Studies 1984: 99–102).

Central control of banking, efficient central co-ordination of planning, and a good supply of up-to-date production data enabled workers' control to perform very well in many cases. The engineering and port industries showed particularly good results and there was some evidence of a widespread improvement in motivation and performance (Schnytzer 1982: 77–83; 116–24). The living standards of the workforce, although low by general European standards, were maintained and in some cases improved as a direct result of the redistribution of incomes, and the combination of modest material success with some genuine devolution of economic power was quite popular. The popularity of the new system was welcome to the Party leadership but the hegemonic control of every aspect of life, amounting literally to totalitarianism, was the real reason why the Party was able to live comfortably with such a significant devolution of power.

### Policies for cultural hegemony

The Party's responsibilities are described officially as keeping Albania socialist, ensuring that the proletariat controls every aspect of development, and defending the proletariat against internal and external exploitation (Institute of Marxist-Leninist Studies 1971: 658–75). There is a formal constitutional democracy with local and national elections in which there are often several candidates, although they have to be from one of the Party organizations. But it is not through this formal political system but through Party control of justice, education, and culture that hegemony is strengthened and maintained.

Natural justice and constitutional law put brakes on planning and propaganda in liberal democracies, which would be unacceptable and considered unnecessary in a post-revolutionary state. Therefore Albanian law asserts the leadership of the Party over the judiciary.

It says that the idea of the absolute independence of the courts of justice is simply a constitutional device for concealing their real class character in bourgeois states. In a socialist post-revolutionary state, they must be subject to the Party.

The educational system is seen as vital background for the development of the post-revolutionary state; education is aimed at preparing the younger generation to take an active part in the construction of a new socialist society. Enver Hoxha said:

> The teacher's task is to impart to young men and women sound
> scientific knowledge, to give them professional skill and the
> correct attitude towards work, to inculcate the Marxist-Leninist
> world outlook, and to imbue them with the spirit of socialist
> patriotism and socialist internationalism, thus ensuring their
> moral, physical and cultural education.
>
> (Hoxha quoted in Ash: 1974: 224)

There is a continuing movement to involve school in what is seen as the revolutionary struggle of daily life. The aim is to avoid schools becoming detached from the rest of the community:

> [Schools] do not merely provide additional personnel for the
> government and planning departments, but help to create a people
> equipped with the knowledge and understanding of the science of
> Marxist-Leninism to play their full role as socialist citizens.
> Learning and education should not be considered as a means of
> profit, like in bourgeois countries, but as a powerful weapon in
> the hand of new man in a socialist society, to promote socialist
> culture.
>
> (Alia 1986: 37)

In the cause of revolutionary culture, artists, sculptors, poets, writers, film-makers, and playwrights are constantly sent the length and breadth of Albania to seek out new talent, to learn from the proletariat, and to see what the proletariat wants from them:

> There will be no artistic freedom. Freedom for whom? For
> individuals who have detached themselves from the masses and
> by selfish pursuit of their own subjective interests, consciously or
> unconsciously collaborating with the class enemies of the
> proletariat. Freedom for whom? For the exploiters and the
> hireling scribblers who serve the interests of their masters by
> ascribing the vileness of the capitalist system to the plight of man
> as such – distracting from the class struggle. No! True creativity
> is by the great working people who alone are capable of changing
> society and changing themselves in the process. This must be

recorded, written about, and form the basis of socialist art and culture in general.

(Xholi 1985: 11–12)

## Conclusion

The Party of Labour of Albania and its leadership have engineered their own legitimacy by an intense programme of propaganda and by control of justice, education, culture, and practically every other aspect of Albanian life. Mineral wealth and a population of fewer than three million have kept Albania free from foreign influence. China and Albania ended their close relationship soon after the death of Mao Zedong, but Albania now has less need of the aid given first by the Soviets and later by the Chinese. A modest but adequate living standard for most Albanians is now possible without foreign subvention.

Enver Hoxha died in 1985 and was succeeded as leader by the uncharismatic Ramiz Alia who has started to upset the delicate balance of worker control and Party power. Attempts to bully the rural population into further collectivization resulted in widespread alienation from the Party and the leadership. The caution of the ageing leadership, who are reluctant to force the pace of cultural, economic, and political change (which Hoxha never feared to do), is alienating the youth. The clumsy administration of Party control has led to complaints that *Drejtim Unik* (single guidance by the Party) which replaced *Udheheqjeunike* (one-person leadership) is a dead hand of restraint. Hegemony needs constant reinforcement and it cannot survive inertia; otherwise, legitimacy, the basis of hegemony, starts to crumble away.

## Power through legitimacy: summary table

*Philosophy of power through legitimacy*

The unequal distribution of power within and between states is accepted as a framework for cultural and economic activity by nearly everybody. This gives it LEGITIMACY. The power of the elite who hold it is pervasive; it is HEGEMONIC.

HEGEL       The first philosopher to discuss the hegemony of the ruling class.

GRAMSCI     Marxist interpretation of the hegemony of the bourgeoisie.

GALBRAITH   A modern interpretation of the hegemony of the technostructure.

Hegel, Gramsci, and Galbraith were primarily concerned with power within the state, but the distribution of power between states is also hegemonic with the capitalist core dominating the capitalist periphery; this is the system of NEO-IMPERIALISM.

Case-studies relating GEOGRAPHY to POWER THROUGH LEGITIMACY
1. ETHIOPIA
   Illustrates neo-imperialism.
2. HEGEMONY AND TOWN PLANNING
   Planners are the agents of the hegemonic state – Sennett offers an anarchist alternative which would still meet the demands of elite hegemony.
3. HIP HOP
   An illustration of a challenge to cultural hegemony.
4. ALBANIA
   Hegemony without capitalism.

*Economic and cultural geography is heavily influenced by the power through legitimacy which results in hegemony within and between states – hegemony would be impossible without the tacit consent of legitimation. The unequal distribution of legitimated power profoundly affects the geography of the world.*

# General conclusion

Geography is a metaphor of politics. Decisions in politics are taken within a total environment; they are not just taken within the political environment of the political scientist, nor within the economic environment of the economist, nor again within the social environment of the sociologist. The geographer's map of salient features has more meaning for the politician than narrower models in social science. Yet geographers have slipped from the political arena and geography seems, if anything, less politically relevant nowadays than other social sciences and humanities. The contention in this book has been that this has happened because geographers have lost touch with political thought and are, therefore, failing to understand the main sources of power in the modern world. If you do not know where the power is, you are at a severe disadvantage in the tough competition for influence. Perhaps by associating geographical models with political power, this book will help to empower geographers once again.

Five sources of power are identified in this book: might, right, nationhood, legality, and legitimacy. Together, it is argued, these explain the phenomena of political geography and the phenomena of human geography as a whole. They are channels of human energy.

Power through might is the power of armed strength and brute force. There is a whole political philosophy built around the power of the strongest able to impose its will, associated with individual and group psychology which explains and models human aggression. The combination of a philosophy of assertion with the psychology of aggression is illustrated in Chapter One by Nazi Germany and the Amazonas, where geopolitical theory is called upon to back up conquest, and by South Africa's relations with Namibia, where theology backs up a policy of assertion and exploitation to supplement the income of a powerful minority. Chapter One, 'Power through Might', concludes with a look at the geography of the battlefield where the conquest of territory through brute force and individual and group aggression is the consuming purpose of everyone involved.

Overall, 'might' is a source of power easy to overlook because it seems so obvious. Sometimes, as on the battlefield, it is a consequence of a power-struggle of a different sort. Often, as in Nazi Germany, the Amazonas, and Namibia, it is the true origin of political activity with overwhelming consequences for human geography.

Power through right is power based on emotional strength and the force of belief. 'Right' as a source of power has a less clear pedigree in political thought than 'might'. Chapter Two, 'Power through Right', explains the relevant political philosophy; it starts with an attempt to understand the idea of ownership through the nature–nurture debate but finds there are no clear solutions. Instead, an explanation of the source of the emotional attachment to territory felt by many individuals and groups is to be found in Piaget's and others' studies of children's development and, in particular, children's emotional attachment to the world around them. There are important insights to be found too in the study of primitive people's feelings of being part of the land; in fact, the political philosophy of the North American Indian chief proves more relevant than that of Rousseau or Locke. Western philosophy comes into its own, however, to explain why the state is such an important focus for emotional attachment; Anderson's 'imagined communities' provide a particularly useful framework for understanding the development of statism from the broader realms of the medieval world. The state soon became the focus for people denied a sense of belonging, as the old hierarchies were discredited and seemed irrelevant to the needs of Europe's developing capitalism.

The case-studies in Chapter Two illustrate different aspects of power through right. In the Holy Land, two ethnic groups claim the right to a state on the same territory. In the Falklands, two existing states fought a war because they claim sovereignty over the same islands. Grenada claimed the right of self-determination in a part of the world where that was unattainable. And, in Northern Ireland, the right to join a neighbouring state is claimed by a large minority, and the province is consequently torn by internal disorder.

Overall, 'right' is a crucial source of power, fixed in human emotions – and equally in ethnic or national emotions – and can only be shifted temporarily by force of arms. The claim of a right to territory is a deeply rooted cause of political activity and such claims are therefore important facts of political geography.

Power through nationhood is power based on the strength of the state which creates the affinities and transactions of nationhood, which is similar in character to ethnicity but coincides with the territory of the state. When the state creates a nation, the result is a particularly powerful entity, a nation-state with internal cohesion necessary for maintaining order and external status to ensure its standing in a statist world. There are certain factors which make it more or less likely that a state will turn into a nation-state, and these are illustrated in the case-studies of Chapter Three, 'Power through Nationhood'.

The case of Norman England illustrates the unplanned early development of the English nation by the embryonic Anglo-Norman state. The case of Guinea in West Africa shows the importance of state integration in the formation of nationhood. The pre-colonial nation-state was succeeded by a period of colonial disintegration. This was followed by a quarter of a century of conscious nation-building, through policies of integration by Sekou Touré which may have enabled the Guinean nation-state to survive the potential disintegration of exposure to the full rigours of neo-imperialism under Lansana Conté in the late 1980s.

Overall, 'nationhood' is a vital source of power for the state. Although there are important subnational and supranational levels of power, the state is the strongest at present. This has been the case for several centuries and will probably be the same for many years to come.

Power through legality is power also based in part on the strength of the internationally recognized status of the state, but it complements power through nationhood in so far as it is concerned with boundaries and territorial definition and not with affinities and transactions within the state. At the same time, it also emphasizes the impact of the internal boundary on political entities below the level of the state. To understand the nature of the legal boundary, the theoretical approach of Tagil and others has provided geographers with a framework of thought, capable of great insight, into the nature of boundaries. This is especially so when it is combined with contributions from political science, especially political conflict theory, and from geography, especially models of delimitation and allocation and of boundaries as barriers. These approaches are illustrated by the case-studies in Chapter Four, 'Power through Legality'.

Each case-study highlights a different aspect of the impact of the legal status of a territorial boundary on the distribution of power. The Sino–Soviet international boundary is one which has been the focus of conflict, originally between two great empires and later between the world's two leading Communist states. The boundary between Britain and the Channel Islands is a kind of federal boundary which shows how a boundary below the level of the state can be used to benefit the political entities on either side of it. The case of the Western Sahara is an example of the impact of boundaries on politics; an artificially bounded Spanish colony became the focus of Saharawi nationalism, but it was allowed to be split by the pseudo-irredentism of its neighbours, Mauritania and Morocco, which resulted in a drawn-out and costly war. Finally, the case of Jerusalem examines the cultural, economic and social consequences of the formal disappearance of the international partition of a city.

Overall, 'legality' represents the formal distribution of power. Just as a constitution, along with internal administrative and political boundaries, legally distributes political power within the state, so international boundaries are at the forefront of international law in the legal definition of the world system of states. International boundaries are the legal territorial barriers to sovereign state power.

Power through legitimacy is power based on the hegemonic organization of the world economy and of culture, politics, and society within states and cities. The philosophy of hegemony from Hegel to Galbraith is traced in Chapter Five and is then illustrated by the model of neo-imperialism and the case of agribusiness.

The first case-study in Chapter Five, 'Power through Legitimacy', illustrates the application of hegemony at the world level to one state, Ethiopia. This is followed by two cases illustrating the operation of hegemony within the state, on town planning and on the culture of the inner city. Finally, the case of Albania provides an example of hegemony following a Marxist revolution.

Overall, 'legitimacy' represents the distribution of power which cannot be explained through force of arms, emotion, the nation-state or legal arrangements. On the contrary, power through legitimacy is often hidden and easy to miss at first sight, which is why it is accepted by the bulk of the population – legitimated by them – despite the surrender of power to a small minority that it entails.

The study of legitimated power concludes the range of five sources of power of which geographers should be aware. They should be aware of them and understand them so as to improve and develop the study of human geography in general and political geography in particular. If they seek influence for themselves, then they must come to grips with political power.

**General summary table**

Geographers should seek the centre stage in politics, because they analyse, model, study, and have a lot to contribute to the world of political decisions and action. In order to be in a position to do this, they must:

1.  understand the philosophy behind the political decisions that matter, the ones that affect people's lives, the ones to do with power;
2.  understand the connections between geography and the use of different types of political power.

This book has attempted to undertake the tasks involved in both 1 and 2 above.

Power is divided up into five types, derived from five sources:

1.  MIGHT          Aggression and conquest.

2.  RIGHT          The mobilization of people who feel they have a right to territory.

3.  NATIONHOOD     The unique power of the nation-state.

4.  LEGALITY       The legal distribution of power and territory.

5.  LEGITIMACY     The reality of hegemonic power within and between states.

Each of these sources of power has an associated political philosophy. Each political philosophy is the basis for political decision-making and political action. *The models used for this decision-making and action are geographical models of the real world.* It is the application of political philosophy to geographical models with its wide variety of real-world outcomes that constitutes the relationship between geography and power.

*Political decisions are made using GEOGRAPHICAL MODELS. If geographers were to learn more about these decisions, they could influence them and improve them.*

# References

Adamolekun, L. (1976) *Sekou Touré's Guinea*, London: Methuen.
*Africa Events* (1988) 'Dicing with students', *Africa Events*, February 1988: 17.
Alia, R. (1986) Report to the 9th Congress of the Party of Labour of Albania on the Activity of the Central Committee of the Party of Labour and the Tasks for the Future, Tirana: 8 Nentori.
Allport, G. W. (1955) *The Nature of Prejudice*, Boston: Beacon.
Allport, G. W. (1960) *Personality and Social Encounter: Selected Essays*, Boston: Beacon.
Almond, G. A. and Verba, S. (1963) *The Civic Culture: Political Attitudes and Democracy in Five Nations*, Princeton: Princeton University Press.
Amin, S. (1976) *Unequal Development: an Essay on the Social Formations of Popular Capitalism*, Brighton: Harvester.
Amin, S. (1977) *Imperialism and Unequal Development*, Brighton: Harvester.
Amnesty International (1985) 'Morocco and Western Sahara' in Amnesty International 1984 Report, London: Amnesty International.
Anderson, B. (1983) *Imagined Communities: Reflections on the Origins and Spread of Nationalism*, London: Verso.
Andramirado, S. (1987) 'Conté face au défi intérieur', *Jeune Afrique*, 16 December 1987: 28–30.
Arendt, H. (1967) *The Origins of Totalitarianism*, London: Allen & Unwin.
Arendt, H. (1970) *On Violence*, Harmondsworth: Penguin.
Aron, R. (1966) *Peace and War: a Theory of International Relations*, New York: Doubleday.
Ash, W. (1974) *Pickaxe and Rifle*, London: Howard Baker.
Auerbach, E. (1953) *Mimesis: the Representation of Reality in Western Literature*, Princeton: Princeton University Press.
Augustine, Saint, Bishop of Hippo (1961) *The Confessions*, Harmondsworth: Penguin.
Avineri, S. (1972) *Hegel's Theory of the Modern State*, Cambridge: Cambridge University Press.
Balési, C. J. (1976) 'From Adversaries to Comrades-in-Arms: West

Africans and the French Military, 1885–1919', unpublished Ph.D thesis, University of Illinois.

Barrett, A. (1982) 'Iron Britannia', *New Left Review* 134.

Barritt, D. P. and Carter, C. F. (1962) *The Northern Ireland Problem: a Study in Group Relations*, Oxford: Oxford University Press.

Barry, M. A. (1988) 'Guinée: révolte contre la vie chère', *Jeune Afrique*, 20 January 1988: 11–13.

Ben-Gurion, D. (1947) 'Minutes of Evidence', Jewish Case before the Anglo-American Committee of Inquiry on Palestine, Jerusalem: Jewish Agency.

Benvenisti, M. (1976) *Jerusalem: the Torn City*, Jerusalem: Isratypest.

Bodin, J. (1961) *Six Livres de la République*, Aalen: Scientia Verlag.

Bradley, K. and Gelb, A. (1983) *Co-operation at Work: the Mondragon Experience*, London: Heinemann.

Bramwell, W. M. and Slowe, P. M. (1986) 'The costly first home: an aspect of housing shortage in two West Sussex towns', *South Hampshire Geographer* 18: 2–8.

Brewer, A. (1980) 'Introduction' in Brewer, A. (ed.), *Marxist Theories of Imperialism*, London: Routledge & Kegan Paul.

Buchanan, R. H. (1982) 'Plantation and colonization: the historical background', in F. W. Boal and J. N. H. Douglas (eds), *Geographical Perspectives on the Northern Ireland Problem*, London: Academic Press.

Bureau Economique du CMRN (1986) *Revue Economique et Financière*, Conakry: Bureau Economique du CMRN.

Bureau Economique du CMRN (1987) *Revue Economique et Financière*, Conakry: Bureau Economique du CMRN.

Cabral, A. (1973) *Return to the source: selected speeches*, New York: Monthly Review Press.

Calvert, P. C. (1983) 'Sovereignty and the Falklands Crisis', *International Affairs* 59: 405–13.

Calvin, J. (1980) *The Institutes of the Christian Religion*, London: Westminster Press.

Carr, W. (1978) *Hitler: a Study in Personality and Politics*, London: Arnold.

Cassese, A. (1986) *International Law in a Divided World*, Oxford: Clarendon.

Cheveau-Loquay, A. (1987) 'La Guinée: va-t-elle continuer a négliger son agriculture?', *Politique Africaine* 25: 120–6.

Child, J. (1985) *Geopolitics and Conflict in South America: Quarrels among Neighbors*, New York: Praeger.

Chole, E. (1987) 'Constraints to industrial development in Ethiopia', *Mondes en Développement* 58: 15–35.

Clausewitz, C. von (1968) *On War*, Harmondsworth: Penguin.

Cobban, A. (1929) *Edmund Burke and the Revolt against the Eighteenth Century*, London: Allen & Unwin.

Colenutt, R. (1988) 'Public Participation in Planning', unpublished working paper for Development Industry Study Group of the Labour Finance and Industry Group.

## References

Connor, W. (1972) 'Nation-building or nation-destroying', *World Politics* 24: 319–55.

Conté, L. (1988) 'The investors who come here are not serious investors', *EEC Courier* 108: 23–6.

Coulborn, R. (1959) *The Origin of Civilized Societies*, Princeton: Princeton University Press.

Crowder, M. (1968a) *West Africa Under Colonial Rule*, London: Hutchinson.

Crowder, M. (1968b) 'West Africa and the 1914–1918 war', *Bulletin de l'Institut Fondémental de l'Afrique Noire* 30: 227–47.

Darby, J. (1986) *Intimidation and Control of Conflict in Northern Ireland*, Dublin: Gill & Macmillan.

Darwin, C. (1972) *The Origin of the Species*, London: Dent.

Deutsch, K. (1966) *The Nerves of Government: Models of Political Communication*, London: Collier-Macmillan.

Deutsch, K. (1979) *National Integration: a Summary of some Concepts and Research*, New York: Free Press.

Deutscher, I. (1966) *Stalin*, Harmondsworth: Penguin.

Diallo, T. (1971) 'Les institutions politiques du Futa Djallon au XIXe siècle', unpublished Ph.D thesis, Université de Paris (Sorbonne).

Diop, M. (1985) *Histoire des Classes Sociales dans l'Afrique de l'Ouest*, Paris: L'Harmattan.

Douglas, D. C. (1969) *William the Conqueror: the Norman Impact on England*, London: Methuen.

Douglas, D. C. and Greenaway, G. W. (1953) *English Historical Documents, vol. II, 1042–1189*, London: Eyre & Spottiswoode.

Duchacek, I. D. (1986) *The Territorial Dimension of Politics: within, among and across Nations*, Boulder: Westview.

Dupuch, C. (1917) 'Essai sur l'Empire Réligieuse chez les Peulh du Fouta Djallon', Paris: Comité d'Etudes Historiques et Scientifiques de l'Afrique Occidentale Française.

Ebeling, G. (1970) *Luther: an Introduction to his Thought*, London: Collins.

Eliade, M. (1973) *Australian Religions: an Introduction*, Ithaca: Cornell University Press.

Fanon, F. (1969) *The Wretched of the Earth*, Harmondsworth: Penguin.

Fermi, L. (1961) *Mussolini*, Chicago: University of Chicago Press.

Fest, J. C. (1970) *The Face of the Third Reich*, London: Weidenfeld & Nicolson.

Fiori, G. (1970) *Antonio Gramsci: Life of a Revolutionary*, New York: Schocken.

Foigny, G. de (1952) 'Terra Australis Incognita', in G. Negley and J. M. Patrick (eds), *The Quest for Utopia: an Anthology of Imaginary Societies*, New York: Schuman.

Franck, T. (1987) 'Theory and practice of decolonisation', in R. Lawless and L. Monahan (eds), *War and Refugees: the Western Sahara Conflict*, London: Pinter.

Frank, A. G. (1979) 'Third World Agriculture and Agribusiness',

University of East Anglia Development Studies Discussion Paper 31.

Galbraith, J. K. (1967) *The New Industrial State*, London: Deutsch.

General Synod of the Dutch Reformed Church (1966) *Human Relations in South Africa: Report of the Committee on Current Affairs*, Johannesburg: Information Bureau of the Dutch Reformed Church.

Ghose, A. K. (1985) 'Transforming feudal agriculture: agrarian change in Ethiopia', *Journal of Development Studies* 22: 127–49.

Gibbs, P. (1922) 'The cemeteries of the Salient', *Ypres Times* 1, 7: 194–9.

Gibbs, P. and Waggett, E. (1922) 'What's the use of the League?' *Ypres Times* 1, 8: 177–8.

Giliomee, H. (1979) 'The development of the Afrikaner's self-concept', in A.R. Hare (ed.), *South Africa: Sociological Analyses*, Cape Town: Oxford University Press.

Gilmore, W. C. (1984) *The Grenada Intervention: Analysis and Documentation*, London: Mansell.

Gilroy, P. (1987) *There Ain't No Black in the Union Jack: the Cultural Politics of Race and Nation*, London: Hutchinson.

Ginsburgs, F. and Pinkele, F. A. (1978) *The Sino–Soviet Territorial Dispute 1949–64*, New York: Praeger.

Goerg, O. (1986) *Commerce et Colonisation en Guinée (1850–1913)*, Paris: L'Harmattan.

Golbery do Couta e Silva (1957) *Aspectos Geopoliticos do Brasil*, Rio de Janeiro: Biblioteca do Exercito.

Gramsci (1971) *Selections from Prison Notebooks*, Q. Hoare and G. Nowell Smith (eds), London: Lawrence & Wishart.

Gregor, A. J. (1969) *The Ideology of Fascism: the Rationale of Totalitarianism*, London: Collier-Macmillan.

Grindea, M. (1982) *The Holy City in Literature*, London: Kahn & Averill.

Griver, S. (1986) 'Cowboy town or holy city', *Jewish Chronicle*, 6 June 1986: 25.

Haggett, P. (1965) *Locational Analysis in Human Geography*, London: Arnold.

Halasa, M. (1988a) 'The life and times of KRS 1', *Soul Underground*, 9 June 1988: 23–4.

Halasa, M. (1988b) 'Rap Attack!' in *Cut*, October 1988: 12–15.

Halliday, F. and Molyneux, M. (1981) *The Ethiopian Revolution*, London: Verso.

Hampshire, S. (1951) *Spinoza*, Harmondsworth: Penguin.

Hancock, G. (1985) *Ethiopia: the Challenge of Hunger*, London: Gollancz.

Hansard Report (1980) Report of the Proceedings of the House of Commons, 2 December 1980, *Hansard*, 995: 128–34.

Harvey, D. (1973) *Social Justice and the City*, London: Arnold.

Hegel, G. W. F. (1942) *Philosophy of Right*, Oxford: Oxford University Press.

Hepple, L. (1986) 'Geopolitics, generals, and the state in Brazil', *Political Geography Quarterly* 5 (supplement): 879–90.

Hitler, A. (1969) *Mein Kampf*, London: Hutchinson.

Hobbes, T. (1949) *De Cive*, New York: Appleton.

Hobbes, T. (1968) *Leviathan*, Harmondsworth: Penguin.

Hodges, T. (1983) *Western Sahara: the Roots of a Desert War*, Westport, Lawrence Hill.

Hoffmann, S. (1981) 'States and the morality of war', *Political Theory* 9: 149–72.

Hopewell, J. F. (1958) 'Muslim Penetration into French Guinea, Sierra Leone, and Liberia before 1850', unpublished Ph.D thesis, Columbia University.

House, J. W. (1982) *Frontier on the Rio Grande: a Political Geography of Development and Social Deprivation*, Oxford: Clarendon.

Howard, M. (1978) *War and the Nation-State*, Oxford: Clarendon.

Hoxha, E. (1975) *Selected Works*, Tirana: 8 Nentori.

Institute of Marxist-Leninist Studies (1971) *History of the Party of Labour of Albania*, Tirana: Naim Frashëri.

Institute of Marxist-Leninist Studies (1984) *PPSH Scientific Conference on the Marxist-Leninist Theoretical Thinking of the Party of Labour of Albania and Comrade Enver Hoxha*, Tirana: 8 Nentori.

International Labour Office (1977) *Labour and Discrimination in Namibia*, Geneva, International Labour Office.

Jinadu, L. A. (1978) 'Some African theorists of culture and modernization: Fanon, Cabral, and some others', *African Studies Review* 21: 121–38.

Johnson, J. T. (1984) *Can Modern War be Just?*, Newhaven: Yale University Press.

Kamil, L. (1986) *Fuelling the Fire of US Policy: the Western Sahara*, Oxford: Oxford University Press.

Katjavivi, F. H. (1988) *A History of Resistance in Namibia*, London: Currey.

Kelly, P. L. (1984) 'Geopolitical themes in the writings of General Carlos de Meira Mattos of Brazil', *Journal of Latin American Studies* 16: 439–61.

Klare, M. (1983) 'The Reagan Doctrine', *New Statesman*, 4 November 1983: 10.

Klineberg, O. (1974) *Race as News: Two General Studies on Attitude Change*, London: HMSO.

Kluckhohn, C. K. M. (1950) *Mirror for Man: the Relation of Anthropology to Modern Life*, London: Harrap.

Krejci, J. and Velimsky, V. (1981) *Ethnic and Political Nations in Europe*, London: Croom Helm.

Kutcher, A. (1973) *The New Jerusalem: Planning and Politics*, London: Thames and Hudson.

Laffin, J. (1979) *The Israeli Mind*, London: Cassell.

Latin American Bureau (1982) *Under the Eagle*, London: Latin American Bureau.

Latin American Bureau (1984) *Grenada: Whose Freedom?*, London: Latin American Bureau.

Laurus, J. (ed.) (1965) *From Collective Security to Preventive Diplomacy: Readings in International Organization and the Maintenance of Peace*, Chichester: Wiley.

Lenin, V. I. (1948) *Imperialism: the Highest Stage of Capitalism*, London: Lawrence & Wishart.

Locke, J. (1966) *The Second Treatise of Government*, Oxford: Blackwell.

Loyn, H. R. (1962) *Anglo-Saxon England and the Norman Conquest*, London: Longmans Green & Co.

Luard, E. (1986) *War in International Society: a Study in International Sociology*, London: Tauris.

Lukes, S. (1974) *Power: a Radical View*, London: Macmillan.

Machiavelli, N. (1961) *The Prince*, Harmondsworth: Penguin.

Mackinder, H. (1969) 'The geographical pivot of history', in R. F. Kasperson and J. V. Minghi (eds), *The Structure of Political Geography*, London: London University Press.

Manuel, F. E. (1965) 'Towards a psychological history of utopias', in F. E. Manuel (ed.), *Utopias and Utopian Thought*, London: Souvenir Press.

Marchés Tropicaux (1987a) 'Réunion du Groupe Consultatif de la Banque Mondiale', *Marchés Tropicaux*, 3 April 1987: 815.

Marchés Tropicaux (1987b) 'Perspectives pour le secteur privé: la mission du Groupe des Sept à Conakry', *Marchés Tropicaux*, 10 April 1987: 871.

Marchés Tropicaux (1987c) 'Guinée', *Marchés Tropicaux*, 28 August 1987: 2272–3.

Marchés Tropicaux (1987d) 'La protocole d'accord sur le prix de la bauxite entrera en vigueur en janvier 1988', *Marchés Tropicaux*, 4 September 1987: 2460.

Markakis, J. (1987) *National and Class Conflict in the Horn of Africa*, Cambridge: Cambridge University Press.

Markakis, J. and Ayele, N. (1978) *Class and Revolution in Ethiopia*, Nottingham: Spokesman.

Markovitz, I. L. (1977) *Power and Class in Africa*, Englewood: Prentice-Hall.

Marx, K. (1981) *Capital*, Harmondsworth: Penguin.

McAllister, I. (1983) 'Class, religion, denomination, and Protestant politics in Ulster', *Political Studies* 31: 275–83.

McEwen, A. C. (1971) *International Boundaries of East Africa*, Oxford: Clarendon.

McLuhan, T. C. (1971) *Touch the Earth: a Self-Portrait of Indian Experience*, Cambridge: Abacus.

Medvedev, R. (1986) *China and the Superpowers*, Oxford: Blackwell.

Michels, R. (1915) *Political Parties: a Sociological Study of the Oligarchical Tendencies of Modern Democracy*, New York: Jarrold.

Milia, F. (1978) *La Atlantártida: un Espacio Geopolítico*, Buenos Aires: Pleamar.

Mill, J. S. (1975) *On Liberty*, Oxford: Oxford University Press.

Miller, G. A. (1970) *The Psychology of Communication*, Harmondsworth: Penguin.

Ministère du Domain Economique (1970) *Revue du Développement Economique*, Conakry: République de Guinée.

Ministry of Education and Culture under the auspices of the Guinean National Commission for UNESCO (1979) *Cultural Policy in the*

*People's Democratic Republic of Guinea*, Paris: UNESCO.
Mirsky, J. (1988) 'Dalai Lama rejects call to revolt', *The Observer*, 3 April 1988: 27.
Mooney, J. (1973) *The Ghost Dance Religion and Wounded Knee*, New York: Dover.
Negrete, J. G. (1972) 'Esquema para una interpretación geopolítica', *Revista Geográfica*, 7: 27–58.
Nicholson, H. (1952) *King George the Fifth: his Life and Reign*, London: Pan.
Nietzsche, F. (1961) *Thus Spake Zarathustra: a Book for Everyone and No One*, Harmondsworth: Penguin.
Nkrumah, K. (1964) *Conscientism: Philosophy and Ideology for Decolonization and Development*, New York: Monthly Review Press.
Oz, A. (1983) *In the Land of Israel*, London: Fontana.
Panaf (1978) *Sekou Touré*, London: Panaf.
Parker, G. (1985) *Western Geopolitical Thought in the Twentieth Century*, London: Croom Helm.
Parsons, T. (1967) *Sociological Theory and Modern Society*, New York: Free Press.
Person, Y. (1969) 'Samori: une révolution Dyula', unpublished Ph.D thesis, Université de Paris (Sorbonne).
Piaget, J. and Inhelder, B. (1956) *The Child's Conception of Space*, London: Routledge & Kegan Paul.
Plato (1955) *The Republic*, Harmondsworth: Penguin.
Prescott, J. R. V. (1985) *The Maritime Political Boundaries of the World*, London: Methuen.
Prittie, T. (1981) *Whose Jerusalem?*, London: Muller.
Ratzel, F. (1969) 'The laws of the spatial growth of states', in R. F. Kasperson and J. V. Minghi (eds), *The Structure of Political Geography*, London: London University Press.
Ridley, N. (1980) Written answers on the Falkland Islands, 1 August 1980, House of Commons, *Hansard* 989: 855–6.
Riegel, J. P. (1988) 'EEC–Guinea co-operation', *EEC Courier* 108: 30–3.
Rivière, C. (1971) *Mutations Sociales en Guinée*, Paris: Marcel Rivière.
Robinson, T. W. (1970) *The Sino–Soviet Border Dispute: Background, Development and the March 1969 Clashes*, Santa Monica: Rand Corporation.
Romann, M. (1981) 'Jews and Arabs in Jerusalem', *Jerusalem Quarterly* 19: 23–46.
Rosenblum, N. L. (1978) *Bentham's Theory of the Modern State*, Cambridge: Harvard University Press.
Rousseau, J.-J. (1968a) 'The Social Contract', in J.-J. Rousseau, *The Social Contract and Discourses*, London: Dent.
Rousseau, J.-J. (1968b) 'A discourse on the origin of inequality', in J.-J. Rousseau, *The Social Contract and Discourses*, London: Dent.
Royal Commission on the Constitution (1975) *The Kilbrandon Report*, London: HMSO.
Schmidt, U. (1983) 'The development of ideological concepts regarding

local organs of power in the Revolutionary People's Republic of Guinea', in G. Brehme and T. Buttner (eds), *African Studies*, Berlin: Akademie-Verlag.

Schnytzer, A. (1982) *Stalinist Economic Strategy in Practice: the Case of Albania*, Oxford: Oxford University Press.

Selcher, W. A. (1977) 'The National Security Doctrine and policies of the Brazilian government', *Parameters: Journal of the US Army War College*, 7: 10–24.

Sennett, R. (1973) *The Uses of Disorder: Personal Identity and City Life*, Harmondsworth: Penguin.

Shaw, M. (1986) *Title to Territory in Africa: International Legal Issues*, Oxford: Clarendon.

Simon, R. (1982) *Gramsci's Political Thought: an Introduction*, London: Lawrence & Wishart.

Sivan, E. (1986) 'Language and nation: the case of Arabic', in J. Alpher (ed.), *Nationalism and Modernity: a Mediterranean Perspective*, New York: Praeger.

Slowe, P. M. (1988) 'Rebuilding an African nation: Guinea from the fifties to the nineties', Paper for the Political Studies Association 1988 Annual Conference.

Slowe, P. M. and Woods, R. H. (1986) *Fields of Death: Battle Scenes of the First World War*, London: Hale.

Slowe, P. M. and Woods, R. H. (1988) *Battlefield Berlin: Siege, Surrender, and Occupation*, London: Hale.

Smith, A. D. (1983) *State and Nation in the Third World*, Brighton: Harvester.

Smith, A. D. (1986) *The Ethnic Origins of Nations*, Oxford: Blackwell.

Snowdon, B. (1985) 'The political economy of the Ethiopian Famine', *National Westminster Bank Quarterly Review*, November 1985: 41–55.

Soja, E. J. (1968) 'Communication and territorial integration in East Africa', *East Lakes Geographer* 4: 39–57.

Soudan, F. and Sada, H. (1988) 'Sahara: les enjeux du référendum', *Jeune Afrique*, 1452: 22–7.

States Housing Authority (1982) *Housing (Control of Occupation) (Guernsey) Law*, St Peter Port: States Housing Authority.

Steiner, G. (1971) *In Bluebeard's Castle*, London: Collins.

Stepan, A. (1971) *The Military in Politics: Changing Patterns in Brazil*, Princetown: Princetown University Press.

Storr, W. (1925) Address by Canon Storr, *Ypres Times* 3, 1: 3–4.

Strachey, A. (1957) *The Unconscious Motives of War: a Psycho-Analytical Contribution*, London: Allen & Unwin.

Strassoldo, R. (1982) 'Boundaries in sociological theory', in R. Strassoldo and G. delli Zolti (eds), *Co-operation and Conflict in Border Areas*, Milan: Franco Angeli.

Strehlow, T. G. (1978) *Central Australian Religion: Personal Monototemism in a Polytotemic Community*, Bloomington: ANZ Religious Publications.

Summers, A. and Johnson, R. W. (1978) 'World War I conscription and social change in Guinea', *Journal of African History* 19: 25–38.

Swift, J. (1939) *A Tale of a Tub*, Oxford: Blackwell.

Tagil, S. (1970) *Deutschland und die Deutsche Minderheit in Nordschleswig: eine Studie zur deutschen Grenzpolitik 1933–39*, Lund Studies in International History 1.

Tagil, S. et al. (1977) *Studying boundary conflicts: a theoretical framework*, Lund Studies in International History 9.

Terrill, R. (1980) *Mao: a Biography*, New York: Harper & Row.

*The Economist* (1987) 'Jerusalem: vox populi, civitas Dei', *The Economist*, 7 November 1987: 23–6.

*The Economist* (1988) 'The peace habit reaches Africa', *The Economist*, 19 November 1988: 91–2.

Thomas, H. (ed.) (1972) *Jose Antonio Primo de Rivera: Selected Writings*, London: Cape.

Thubten Jigme Norbu (1986) *Tibet is my Country: the Autobiography of Thubten Jigme Norbu, Brother of the Dalai Lama*, London: Wisdom.

*The Times* (1982) 'We are all Falklanders now', *The Times*, 4 April 1982: 10.

Tissier, A. H. le (ed.) (1985) *Berlin Soldier*, Berlin: Berlin Bulletin.

Touré, A. S. (1959) *La Lutte du Parti Démocratique de Guinée pour l'Emancipation Africaine*, Conakry: Bureau Politique National.

Touré, A. S. (1964) *Plan Septennal 1964–71*, Conakry: République de Guinée.

Touré, A. S. (1969) *La Révolution Culturelle*, Conakry: République de Guinée.

Touval, S. (1966) 'Africa's frontiers: reactions to colonial legacy', *International Affairs* 42: 641–54.

Traoré, A. (1988a) 'Guinea: painful rebirth', *EEC Courier* 108: 16–22.

Traoré, A. (1988b) 'Bauxite for the next 500 years', *EEC Courier* 108: 27–8.

Traoré, D. (1962) *Samory Sanglant et Magnifique*, Conakry: République de Guinée.

Travassos, M. (1935) *Projecao Continental do Brasil*, Sao Paulo: Editorial Nacional.

Treitschke, H. von (1963) *Politics*, New York: Harcourt, Brace & World Inc.

Tsien-Hua Tsui (1983) *The Sino–Soviet Border Dispute in the 1970s*, Oakville: Mosaic.

Turner, J. (1986) 'Polisario rebels raise the stakes', *South* 66: 19–20.

Verney, D. V. (1986) *Three Civilizations, Two Cultures, One State: Canada's Political Traditions*, Durham: Duke University Press.

Walzer, M. (1978) *Just and Unjust Wars: a Moral Argument with Historical Illustrations*, London: Allen Lane.

Weigert, H. W. (1942) *Generals and Geographers: the Twilight of Geopolitics*, New York: Books for Libraries Press.

Weiner, M. (1971) 'The Macedonian Syndrome: an historical model of international relations and political development', *World Politics* 23: 665–83.

Wesson, R. and Fleischer, D. V. (1983) *Brazil in Transition*, New York: Praeger.

Whiting, A. S. (1957) *Soviet Policies in China 1917–24*, New York: Columbia University Press.

Xholi, Z. (1985) *For a More Correct Conception of National Culture*, Tirana: 8 Nentori.

Yahuda, M. (1983) *Towards the End of Isolationism: China's Foreign Policy after Mao*, London: Macmillan.

*Ypres Times* (1921a) (editorial) *Ypres Times* 1, 1: 9.

*Ypres Times* (1921b) 'Ypres League news', *Ypres Times* 1, 2: 53–9.

*Ypres Times* (1923) Sheffield Branch Dinner, *Ypres Times* 2, 1: 16–19.

*Ypres Times* (1924) The Junior Division of the Ypres League, *Ypres Times* 2, 3: 70–3.

# Index